Taming the Atom:

Facing the Future with Nuclear Power

To Lynda

Taming the Atom:

Facing the Future with Nuclear Power

Ian Blair

Adam Hilger Ltd, Bristol

British Library Cataloguing in Publication Data

Blair, Ian M.
Taming the Atom: Facing the Future with Nuclear Power
1. Atomic power
I. Title
621.48 TK9145

ISBN 0-85274-414-5
ISBN 0-85274-483-8 (Pbk)

Consultant Editor: **Professor N H Lipman,**
Rutherford and Appleton Laboratories

Published by Adam Hilger Ltd,
Techno House, Redcliffe Way, Bristol BS1 6NX

The Adam Hilger book-publishing imprint is owned by The Institute of Physics

Printed in Great Britain by
J W Arrowsmith Ltd, Bristol

Contents

Preface vii

Acknowledgments x

1 The Mythology of the Atom 1

2 What is Nuclear Power? 8
The Atom and its Nucleus 12
Radioactivity 18
Nuclear Fission 26
Breeding Nuclear Fuel 33
How a reactor works 35
The natural reactor at Oklo 39
The fast reactor 43
Nuclear fusion 46

3 The Nuclear Industry in Profile 48
Uranium mining 52
Isotope enrichment 60
Reactor fuel fabrication 66
Types of reactor 68
Decommissioning redundant stations 77
Transport of spent nuclear fuel 79
Reprocessing the spent fuel 84
Management of waste products 88

4	Nuclear Power in the Energy Scene	100
	Energy in man's development	101
	The impending crisis	105
	The need for energy conservation	112
	The role of nuclear power	118
	Status of the fast reactor programme	126
	Atoms by wire	128
	Other possible sources	133
	The question of economics	144
	The next few decades	148
5	Matters of Public Concern	150
	Biological effects of radiation	152
	Probability and consequences of accidents	163
	Worries about waste disposal	175
	No free lunches	179
	The technological imperative	186
	The centralisation of power	187
	Fears about terrorism	191
	Threats to civil liberties	196
	Proliferation of nuclear weapons	198
6	The Great Nuclear Debate	202
	Depth of public concern	207
	Lack of public knowledge	214
	Differing national techniques	218
	Put it somewhere else!	224
	A question of credibility	227
7	The Real Issue	230
Appendix 1	World Nuclear Reactor Types	237
Appendix 2	Bibliography and Sources of Further Information	246

Preface

I suppose that it can be said of most books that they did not just happen, but rather that there is a story behind them. Certainly that is the case with this one.

Having been involved in the nuclear debate for a number of years, it struck me as rather odd that whereas there were a large number of popular books on the subject, none of them had been written—at least up until then—by people who were currently earning their living by working in the industry. I mentioned this anomaly vociferously on a number of occasions, and so should not have been surprised when one day I received a telephone call from Neville Goodman, who was at that time a Commissioning Editor with Adam Hilger Ltd, the gist of which was that he had heard that I was writing a popular book on nuclear power and please could his firm publish it!

Now of course I was not at that time writing such a book, nor had I any plans to do so, but then I thought that I could not continue castigating others for not doing what I was not prepared to do myself. So, with some trepidation, not ever having written a book before, I decided to accept the challenge.

Next I had to decide what kind of a book I wanted to write, what it should contain and for whom I was writing it. My experience of the nuclear debate had taught me that there was a large number of people who, although they had little or no technical background, felt that the issue of nuclear power was sufficiently important that they ought to get to grips with it so that they could form a personal view. What they needed, therefore, was a straightforward account that was both authoritative and readable to act as a basis. It was such a reader—sometimes euphemistically called 'the intelligent layman'—that I had in mind whilst I was writing my book.

The core of the book had to contain a description of the processes involved in extracting electricity from uranium and how the nuclear industry is organised to do that. On the grounds that one does not build nuclear power stations for fun, but rather because society needs the electricity that they produce, it was necessary to describe the present energy situation and how nuclear power might contribute to its development over the years ahead.

First of all, however, I felt that I had to dispel some of the myths that have accreted to nuclear power over the years so that my discussion could start from a basis of fact; hence the title of my first chapter. Finally, I had to discuss the issues that most concern people about nuclear power and describe how the nuclear debate is conducted in practice, leading to what I believe is the major underlying motivation of that debate. I deal with these topics in the later chapters.

The reader will notice two things about this book. The first is that it is a very personal account: it is my book and not the Atomic Energy Authority's book. I have, of course, received many valuable comments from colleagues at Harwell and elsewhere which have certainly improved my knowledge of those areas of the subject in which it was weak. But at the end of the day the opinions expressed are my own, and do not necessarily correspond with those of my employers, although I hope that they will not dissent from them too strongly.

The second, which follows from the first, is that the style is somewhat idiosyncratic, and can perhaps best be described as anecdotal. I make no apology for that, as I believe that anecdotes, particularly when drawn from life—and I can vouch personally for the historicity of most of them—can illustrate a point much more effectively than any amount of turgid descriptive prose.

I have found writing my first book an enjoyable and instructive experience. I can now only hope that the reader finds it similarly enjoyable and instructive.

Ian Blair

Acknowledgments

I must first of all thank many of my colleagues, too numerous to mention individually, who took the trouble to read and comment upon the earlier drafts of this book. They pointed out several errors and suggested many improvements. The time they must have spent went far beyond the call of friendship or duty, and I am extremely grateful to them.

Next I must thank the United Kingdom Atomic Energy Authority for its permission to publish this book. Although it is my own book written in my own time, because it concerns work which I do for the Authority as its employee, it was naturally interested in what I had to say. I must also thank the UKAEA and the following for permission to use photographs for which they hold the copyright: Australian Information Service, British Nuclear Fuels Ltd, Cavendish Laboratory, Chicago Tribune, Carolina Light and Power Co., Jersey Evening Post, Lotte Meitner-Graf ARPS, Ontario Hydro and Sandia National Laboratories.

Finally I must thank Mrs Enid Jones, whose professional skill turned my somewhat illegible handwriting into the final typescript.

CHAPTER 1

The Mythology of the Atom

The Nuclear Age began with a bang. More precisely, it began with two bangs. One over Hiroshima and the other over Nagasaki—two unfortunate Japanese cities that in a fraction of a second ceased to exist. From that moment, the world knew that it had in its hands a force of terrifying proportions. There were immediate international consequences. The Second World War, which had raged for six years and was expected to continue for many more, rapidly came to an end. An age of suspicion and mistrust—the Cold War—between the victorious former allies began.

Gradually the memory of the horrors of war faded. The main concern of most people in those years was the rebuilding of their lives and societies following the ravages of the worst war that the world had ever known. The Bomb existed, and no amount of wishful thinking would make it go away.

It had also become clear that this fearsome power could be harnessed for good. That which threatened man's destruction could also promise his salvation. The raw material of the Bomb—uranium—could alternatively be used to fuel a power station. It is by no means a rare element—it is more abundant than tin, for example—and one ton of it has the energy potential of more than a million tons of coal. A cheap and abundant supply of power was what had enabled men to rise up from the grinding poverty of an agrarian society. Here was abundant power indeed—enough to drive human advancement for centuries to come!

It is difficult today to recall the euphoria of those early days, when many of the best young brains of the time rushed to enroll in this dynamic new industry. The power for the New Age would come from the atom and they would have a hand in developing it.

Atomic scientists were revered above all other men. They bestrode the world like colossi. Their views were sought not just on atomic energy, about which they could justly claim to be experts, but on all matters from philosophy to politics and religion. Albert Einstein, the Supreme Being of the New Pantheon, was even offered the presidency of the new state of Israel. Very wisely, he refused. Such humility was rare, however, for it is difficult for ordinary mortals not to be profoundly affected psychologically by such adulation. A new cult had been born. There was no god but the atom and the atomic scientists were its prophets.

Such then was the mood of public optimism which sprang up in reaction to the horrors of the Bomb. The ghosts of Hiroshima and Nagasaki were not to be so easily laid, however. They had merely been buried in the corporate subconscious. It is otherwise difficult to explain the recent upsurge, some thirty years later, of undoubtedly sincere public concern about the use of civil nuclear power. The major worries that people have—such as the misuse of civil nuclear power for war-like purposes by irresponsible governments or terrorist groups, the escape of radioactivity and the effects of radiation on man—have all been well known from the early days. Why then are such worries surfacing now? Why are we having a nuclear debate today and not when it all started in the late forties? Is it that a generation, which knew only too well the disastrous consequences of misusing nuclear power and reacted only with relief to see swords being beaten into ploughshares, has passed to be replaced by another that has not known the horror at first hand but fears that someday it might? Have the ghosts returned to prick the sleeping conscience of an exceptionally fortunate generation? Whatever the explicit issues raised in a debate on civil nuclear power, behind it all stands the shadow of the Mushroom Cloud.

In our own day a new atomic cult has arisen. It is a more penitential and escapist cult than the earlier one. Whereas our fathers sought to assuage the guilt of Hiroshima and Nagasaki by beating their swords into ploughshares, the new cult attempts to do so by expurgating the word nuclear from man's vocabulary. All nuclear development, for good or evil, must be abandoned. If that means reducing our standard of living, then so be it. It is the penance we must perform to erase the sins of our fathers from the memory of our race.

Like all effective religions, it has its canon, its priesthood and its demons. Every true believer holds the view that nuclear power is so dangerous that just one slip, one error of judgment by a white-coated technician, will mean the end of the world. The effluent from the nuclear industry is slowly poisoning the Earth, destroying the environment and upsetting the ecology. Nuclear waste will remain dangerous over geological periods of time and no one knows what to do with it. The development of nuclear power will mean that more and more nations will acquire nuclear weapons, thus making a nuclear holocaust inevitable. Nuclear material must be guarded with military-style security to prevent terrorists getting their hands on it and holding society to nuclear blackmail, and to enforce this security all the attributes of a totalitarian state, which would erode our civil liberties and democratic institutions, are required. None of this is true, of course, but no one has yet succeeded in diverting a worshipper from his devotions by pointing out to him that the basic tenets of his faith are unsound.

The priests are more like itinerant friars than classical gods. They bear the lean and hungry look arising from a humble and ascetic lifestyle, and in their eyes is the distant stare of those on whose shoulders rests the conscience of the world. They have an inordinate propensity for martyrdom. The demons are those hard-headed amoral men of the Nuclear Establishment, hell-bent on foisting their technology on an unsuspecting public and operating behind an inpenetrable cloak of official secrecy.

Such then is the mystique that surrounds nuclear power—

an aura stretching from adulation at one extreme to blind terror at the other—and which clouds our judgment and inhibits rational debate. If we are to obtain a balanced view of nuclear power, to appreciate what it can and cannot do, how and to what extent it can contribute to man's development and what its environmental consequences might be, the first thing we must do is to demythologise it. It is for that purpose that I presumed to write this book.

It is a very personal account, and so perhaps I had better say a little about myself and how I came to be involved.

Although I have been working at Harwell since 1964, my career from then until 1973 involved research projects of a very fundamental nature, carried out almost exclusively with groups of scientists from various universities. Such things as hyperon beta decay, nucleon isobar production, off-energy-shell effects and three-body forces and the like. All good stuff and vital to a basic understanding of the laws of nature, but certainly not of immediate practical use for the material betterment of mankind.

For a while I was even involved in the study of Moon rocks, and it was in the course of this project that I first developed a taste for the public platform which hitherto had always filled me with dread and foreboding, if not naked fear. Until then no one, at least amongst the general public, had shown the slightest interest in my research but with the Moon all that suddenly changed. It was topical. Public interest was widespread. I was invited to appear on television and radio and was interviewed by the press. I had a heavy programme of speaking engagements at schools, clubs, organisations and societies of all kinds. After years of working away quietly in the laboratory, this glare of publicity seemed most strange. Gradually the panic of appearing in public subsided and I actually began to enjoy it.

I say I gave lots of talks, but inevitably they evolved into one talk given lots of times, albeit modified to suit the particular audience. The rub came when I was invited back by a society in Oxford to give a *second* talk on the subject, so I had to knuckle down and prepare an entirely new one. Perhaps the most challenging appearance was before a group of about sixty 10–11 year olds at a local junior school. I had invited them to interrupt with questions as we went along, and after about ten minutes the first tentative hands started going up. These soon became a veritable forest, and the questions kept coming for over two hours. It was an exhausting business, but it was all good practice for later, as we shall see. A few days later I received a folder from the school, containing individual letters of thanks from each and every one of them, which I have kept as a treasured possession to this day. The last Moon talk that I gave was at a university department early in 1973. Only about six people turned up. In the next room there was a lecture on environmental issues, and it was filled to capacity. Clearly public interest had shifted markedly.

Indeed, I myself had become very interested in such issues by then, and had begun to read avidly the growing number of publications on the subject. With the advent of the oil crisis I started jumping up and down in a rather agitated manner demanding to know why Harwell was not turning loose its vast scientific expertise into a wider sphere of energy problems including energy conservation and the renewable energy sources. I was scheduled to attend a meeting at the Department of Energy to discuss such matters, but received a 'phone call at home the previous evening and was told, from a rather high level, that I would be having a cold next morning and would therefore be unable to go! As I have always been fascinated by cloak and dagger tactics, I was intrigued rather than offended by such apparently high-handed treatment. On arriving at Harwell the next morning (my fictitious cold having cleared miraculously) I was offered the explanation that courtesy demanded. Unbeknown to me, what had been happening behind the scenes was that negotiations had been going on to set up a unit—ETSU, the Energy Technology Support Unit—at Harwell to do just those things that I had been shouting about. The negotiations had reached a delicate stage—for example, other energy interests had to be convinced that Harwell was the right location for such a unit—and the last thing they wanted at that time was a young fool like me (though they didn't put it quite like that—they were being courteous, remember) hopping around the Department of Energy and stepping, albeit inadvertently, on sensitive corns. Having given me that explanation, and knowing my deep interest in the subject, they could do no other than offer me a job in the new unit, which I accepted gladly. So it was that I became a founder member of ETSU. In fact, I was a member before it was founded, which I suppose qualifies me for some sort of record.

My main interests were to be in the energy conservation field, and I became involved specifically in studies on heat pumps, combined heat and power schemes and the use of reject heat from power stations. Although not directly concerned with the renewable energy sources, we were at that time a small and compact unit and hence I was in close contact with my colleagues who were so concerned during the period over which their studies developed and their ideas matured. I was also instrumental in the setting up and running of a highly successful Summer School on all aspects of energy, which brought me into contact with a wide spectrum of energy interests. It caused me some concern to find that the work in ETSU was to be largely desk bound. We did receive invitations to speak at public meetings and to address various organisations, but such invitations were not encouraged, and certainly not actively sought. This was, and still is, a pity, because it had become clear even by then that the major limitation to the widespread deployment of energy conservation in particular was not technical feasibility or economic viability but the motivation of the energy user to act. On a more personal level, I came to realise how difficult it must be for those who have become accustomed to the limelight, either in the theatre or in politics, to move backstage.

Meanwhile the nuclear debate had begun to hot up, and it became clear to me from a study of more general energy matters that it made an enormous difference to our future energy prospects whether we could or could not assume that nuclear power would make a substantial contribution to that future. It seemed to depend crucially on the nature and magnitude of the environmental impact of nuclear power, and the consequent public acceptance of this relatively new technology. Such was my concern over this basic question that I felt that I had to get involved myself. Consequently, towards the end of 1976, I moved into a small outfit at Harwell called the Nuclear Environment Branch which had been set up some time previously to look into such matters.

So it was that I was to find myself in the front line of the nuclear debate. I brought with me a general background in nuclear matters from my years of working at Harwell (but without the emotional commitment of many of my senior colleagues who had devoted most of their working lives to the development of nuclear technology), a wider background of the more general energy situation from my time in ETSU and a taste for the public platform acquired from my Moon days and from my more recently developed extramural activities in the political sphere.

CHAPTER 2

What is Nuclear Power?

Man has always been a visionary animal. He is perpetually reaching beyond the ordinary and the commonplace into the realm of the extraordinary and the unexpected. It is an attribute which seems to be basic to his nature. It is one of the features which characterises his humanity. Way back in antiquity, and over many centuries, alchemists searched for the 'philosopher's stone'—the magic ingredient that would transform a base metal such as lead into a noble one such as gold. The motivation was greed—another attribute which seems to be basic to human nature. To him who succeeded in the quest there would be riches beyond belief.

As our knowledge of the natural world increased, it became more and more obvious that the quest was an impossible one. We learned that our world was composed of a limited number of 'elements'. These elements could combine with others to form the rich variety of chemical compounds with which our world abounds, but each element consisted of discrete particles—called 'atoms'—which were identical and unique to each element, and which were indestructible and immutable. There was no way in which one element could be transformed into another. Lead was one element, and gold was another. There was no way, therefore, that lead could be transmuted into gold. Such was the received truth of classical nineteenth-century science. Part of the excitement of science, however, is the way that it develops and evolves with time—how the received truth of one generation becomes the approximate truth of the next. In general it is true that one element cannot be transformed into another, but it is not always true.

Sir John Cockroft. Together with Ernest Walton, he was the first man to split the atom by artificial means. Later he became the founder and first Director of the Atomic Energy Research Establishment at Harwell. (© UKAEA.)

Early in the 1930s, two brilliant young scientists—John Cockcroft and Ernest Walton—were working at the Cavendish Laboratory in Cambridge under the direction of the world-famous New Zealand scientist Professor Ernest Rutherford, who was later to be ennobled for his services to science. Their task was to achieve what a century of scientific tradition had insisted was impossible—to 'split the atom', to transform one element of nature into another. In 1932 they succeeded. True, they had only transformed lithium into helium, rather than lead into gold, but it was the principle that was important.

A model of the apparatus with which Cockroft and Walton split the atom, which is on display at Harwell. (© UKAEA.)

Man had learned how to transmute the elements. Yet another barrier had been penetrated, yet another obstacle overcome, in man's quest to understand his world, and with that understanding to fashion it to his use.

Shortly after this momentous discovery, Lord Rutherford was asked whether it would have any practical application. After all, tremendous forces are released in the splitting of an atom. Could not these be used by man for his benefit? Rutherford replied that they could not, for a very simple reason: strong forces were indeed released when splitting an atom, but because it was so tightly bound, even greater forces were needed to split it. It was therefore not possible to get a net production of energy in the process. As to the practical value of being able to convert a useless material into a valuable one, surely that would merely make the useless material itself valuable?—a fact one might have thought would have been obvious to the alchemists of old.

Lord Rutherford. Strictly speaking he could claim to be the first man to split the atom, using alpha particles from a radioactive source. This portrait of him now hangs in the laboratory that bears his name alongside Harwell. (© Cavendish Laboratory, Cambridge.)

In the event, Rutherford turned out to be wrong. In principle he should have been right, but was proved wrong by that time-honoured and endearing trick of science—the exception that proves the rule. His two students would appear to have been more perceptive. John Cockcroft continued with this line of research and eventually became the father of the British nuclear industry, receiving a knighthood for his services. Ernest Walton, with perhaps a more pessimistic view of human nature, foresaw the evil men might do with this powerful force. He opted out of nuclear research and took up a university teaching post in Ireland.

THE ATOM AND ITS NUCLEUS

It is appropriate at this point in our story to review the scientific background of the world of the atom, so that we can better understand the significance of Cockcroft's and Walton's discovery and the consequences to which it was to lead.

Originally, atoms were thought of as small spheres: homogeneous, featureless and identical for a given element, but different from one element to another. They were later shown to be more complicated than that. They consist of two basic parts—the nucleus and the electron cloud.

The nucleus is small and compact and contains the bulk of the mass of the atom—well over 99·9%. It is a composite of two of the fundamental particles of nature—the proton and the neutron. These particles are of roughly the same mass, although the neutron is a tenth of a per cent or so heavier. The proton carries one positive unit of electrical charge whereas the neutron is electrically neutral. The powerful force that binds protons and neutrons together in the nucleus, the 'strong interaction' as it is called, does not distinguish between protons and neutrons. So the force between a proton and another proton is exactly the same as that between a proton and a neutron or between a neutron and another neutron. For this reason, neutrons and protons, the constituents of every atomic nucleus, are sometimes referred to collectively as 'nucleons'.

You may remember from your knowledge of elementary science that like electrical charges repel each other. How is it then that several protons, each carrying one unit of positive charge, can co-exist inside the tiny confines of an atomic nucleus? Why don't the protons repel each other, and cause the nucleus to blow apart? The answer is simple, but highly significant. The nuclear force which binds the protons together, the strong interaction, is very much stronger than the electrical force which is trying to push them apart. In fact it is over ten thousand times stronger.

The strong interaction and the electrical force (or, more accurately, the electromagnetic interaction through which electrical and magnetic forces are connected) are two of the four forces known to exist in nature. The others are the weak interaction, responsible for some forms of radioactive decay, as we shall see later, and gravity. The weak interaction is a million million times weaker than the strong interaction, and gravity is many million times weaker still. One can appreciate the force of gravity by observing the rise and fall of the tides, or the paths of the planets in their orbits. More masochistically, one can feel its strength by dropping a brick on one's toe, or by following Sir Isaac Newton who, so tradition has it, came to appreciate the force of gravity when an apple fell on his head. By comparison, one can perhaps begin to get an inkling of the enormity of the forces which bind together the nucleus of an atom.

Now let us return to the nucleus of the atom and ask how it varies from one element to another. Let us start with the simplest atom, that of the element hydrogen. The nucleus of the hydrogen atom consists of a single proton. The next simplest is helium, whose atomic nucleus consists of two protons and two neutrons; then comes lithium, with three protons and four neutrons; next beryllium with four protons and five neutrons, followed by boron with five protons and six neutrons, and carbon with six protons and six neutrons, ...

We thus see a pattern emerging. The feature which characterises the atom of an element is the number of protons in its nucleus. There may be a few neutrons as well, whose number is roughly similar to the number of protons, but it is the proton number, called the 'atomic number', which defines the element. Thus hydrogen has atomic number one, helium two, lithium three and so on. The series continues up to the

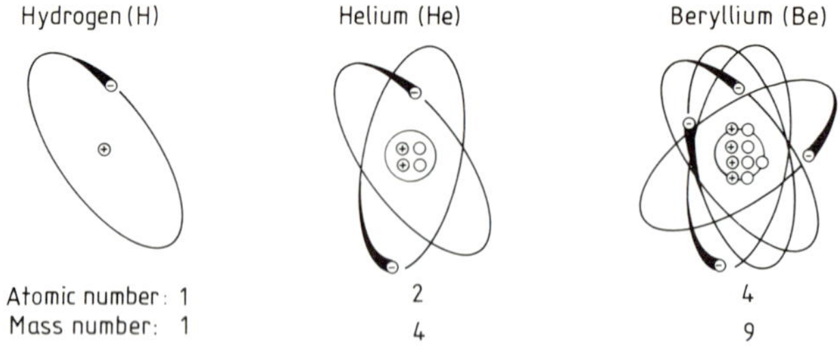

The number of positively charged protons (+) determines the element to which a particular atom belongs. The number of protons plus the number of neutrons (o) determines the mass of the atom. The negatively charged electrons are represented as −.

largest atom in nature, that of uranium, with atomic number 92. Larger atoms can be made artifically, but these do not occur in nature for reasons that we shall see later.

There is no similar regular pattern in the number of neutrons in the nucleus, save that it is roughly the same as the number of protons. In fact most elements exist in nature in different forms, that is with the same number of protons, which characterises the element, but with different numbers of neutrons. Thus there is a form of hydrogen, called deuterium, whose atomic nucleus consists not just of a single proton, but of a proton and a neutron and there is a form of lithium which has three rather than four neutrons associated with its three protons. These different forms are referred to as 'iso-

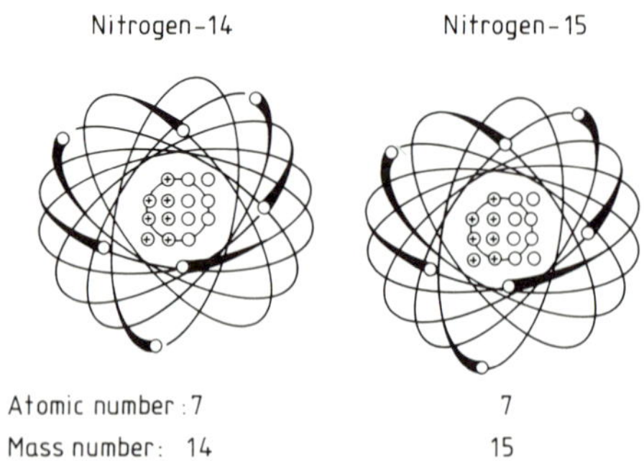

Isotopes of an element differ only in the number of neutrons in the nucleus; they have similar chemical properties but different physical ones.

topes' of the elements. Thus there are two isotopes of hydrogen in nature, the ordinary one with a single proton as its atomic nucleus and deuterium which has a proton and a neutron. Lithium has two isotopes, one with three and the other with four neutrons associated with the three protons in its atomic nucleus.

As the mass of the neutron is very nearly equal to that of the proton, the total number of protons plus neutrons is called the 'atomic mass' of the isotope. Thus lithium has two isotopes of atomic mass six and atomic mass seven, corresponding to the forms with three and four neutrons, respectively, in its nucleus. In a convenient shorthand, these are referred to as lithium-6 and lithium-7.

Most elements occur in nature in several isotopic forms. Typically the number of naturally occurring isotopes of an element is three or four, but there are some, such as iodine, which have only one and others, such as tin, which have as many as ten. The number and relative abundance of the isotopes of a given element in nature are always constant. Thus lithium always exists as 7·5% lithium-6 and 92·5% lithium-7, and iron always exists as 5·8% iron-54, 91·7% iron-56, 2·2% iron-57 and 0·3% iron-58. The reasons for these observed abundancies and their constancy result from the processes involved in the formation of the Earth some four and a half billion years ago—processes which we are only just beginning to understand.

The element uranium, which is soon to become the centrepiece of our story, exists in nature as three isotopes. The most abundant form is uranium-238, which makes up 99·3% of natural uranium. The others are uranium-235, which makes up 0·7%, and uranium-234, which only occurs in trace amounts (0·005%). As with all other elements these ratios are constant and independent of the source of the uranium. There has been one bizarre incident when this was not so, but we shall see the reasons for that later. By far the most important isotope of uranium for the purposes of our story is uranium-235, because this is the only one that can be used as a nuclear fuel. In fact it is the only naturally occurring isotope of any element that can be used as a nuclear fuel, but more about that later.

So far we have dealt only with the nucleus of the atom. What about the rest of it—the electron cloud? The electron is another of the fundamental particles of matter. Its mass is only about 1/2000th of that of a proton and it carries one unit of negative electric charge. The number of electrons in the electron cloud of an atom is exactly equal to the number of protons in the nucleus, and is hence equal to the atomic number. This equality ensures that the atom as a whole is electrically neutral. From schoolday physics we know that opposite electrical charges attract each other, and it is this attractive force that holds the electron cloud round the nucleus.

It is relatively easy to remove one or two electrons from the electron cloud of an atom. Running your comb through your hair, or pulling your shirt over your head will do this. The sparks you see afterwards are caused by the electrons dropping back into their place, from which you had removed them by the vigour of your action. I suppose that technically you could have claimed to have split the atom, and in a pedantic sort of way you might be right. By splitting the atom, however, I hope it has become clear that what is meant is splitting the *nucleus* of an atom. That is a much more difficult proposition, as the forces binding together the nucleus of an atom are some ten thousand times stronger than those binding the electrons in the electron cloud to the nucleus. To attempt to split the atom in the truer sense of splitting the nucleus is therefore not to be recommended by either of the methods above. As we have seen, it was not until 1932 that this was first achieved by Cockcroft and Walton using somewhat more sophisticated apparatus than that described in our homely example.

It is through their electron clouds that atoms combine with other atoms to form the rich variety of chemical compounds with which our world abounds. They do this by sharing or exchanging electrons. Thus the forces which bind atoms together in the basic unit of a chemical compound—the molecule—are electrical in nature. So chemical reactions, in which compounds react to form other compounds, rely on the electromagnetic interaction as their driving force.

The fossil fuels such as coal, oil and gas, so called because they are produced from the remains of long-deceased vegetable matter, are classified chemically as hydrocarbons. This is because they are a mixture of chemical compounds of carbon and hydrogen. When they are burnt, a chemical reaction takes place with the oxygen in the air and the end products of this reaction are carbon dioxide and water. Energy is released in the form of heat, and as this is a chemical reaction the energy is released through the agency of the electromagnetic interaction.

As we saw earlier, this electromagnetic interaction, which binds the electron cloud of an atom and which is the driving force of chemical reactions, is some ten thousand times weaker than the strong interaction which binds the nucleus of an atom together. Therefore nuclear energy, derived from the splitting of the atomic nucleus, is a vastly more concentrated form of energy than that released by chemical processes such as burning, although in both cases the energy manifests itself in the form of heat. One consequence of this is that one gets a lot more energy from a ton of nuclear fuel such as uranium than from a ton of a fuel such as coal or oil. The atomic masses are also important, and as the atomic weight of uranium is some hundred times or so greater than that of the atomic weights of the elements such as carbon and hydrogen which make up the fossil fuels, this gives a further concentration factor of about this magnitude. One might then expect that nuclear energy from uranium will be something like ten thousand times one hundred, i.e. a million times more concentrated than the energy obtained from burning coal; therefore one ton of uranium should give us as much energy as roughly a million tons of coal. In fact, as we shall see later, that does indeed turn out to be the case.

When Cockcroft and Walton first split the atom in 1932 it was thus the first step on the road to an energy source of unbelievable proportions. The significance was not grasped at the time, for in essence Lord Rutherford was right. How could you get more energy out than you needed to put in to split the nucleus? Tremendous forces were at work in the nucleus of the atom. How could they be released to work for man? The answer came with a further discovery of a rather peculiar property of the nucleus of the isotope uranium-235. But that discovery was not made until 1938, and by then the dark clouds of war were gathering over the world. As we shall see, that too was not without its significance.

RADIOACTIVITY

I have gone to some lengths to stress the importance of the work of Cockcroft and Walton. Indeed, splitting the atom was an important and significant event but, as so often happens, and it is a necessary humbling experience for man, nature had done it first. Man learned how to fly in 1903 when the Wright brothers' contraption (I hesitate to call it an aeroplane) lifted off the ground. However, birds had been flying for many millenia. So it was with the splitting of the atom: man first learned how to do it in 1932, but nature again had been doing it for many millenia—some millions of millenia in fact, ever since the world began. That was through the natural phenomenon of radioactivity.

Radioactivity was first discovered, as many things are, almost by accident. One day in 1896 a French scientist called Henri Becquerel put a piece of uranium ore, which he had been examining, away in a drawer. By chance, there were some unexposed photographic plates in the same drawer. Some time later, when Becquerel took out these plates to use them he found that they had become fogged and therefore useless. Had he been a 'normal' person rather than a scientist, he might have cursed volubly and gone out to buy some more plates. Anyone who thinks that scientists are normal people has obviously never worked with them. They have a number of endearing eccentricities, one of which is an uncontrollable excitement and almost Pavlovian reaction to the unexpected event.

Henri Becquerel decided to investigate what had happened

and came to the conclusion that some as yet unknown radiation similar to the x-rays which had recently been discovered, was coming from the ore. There followed twenty years of intense study by the scientific community, which at that time included such famous names as Pierre and Marie Curie and Ernest Rutherford. Surprisingly, this new radiation seemed to be unaffected by temperature, pressure or the chemical state of the element emitting it—surprising because these variables had previously always had a dramatic effect on natural phenomena. Eventually the truth was unearthed. What was happening was a transformation affecting the nucleus of the atom itself. No wonder the usual variables had no effect—at best they can only impinge on the electron cloud of the atom. Nature was transmuting one element into another under man's very nose, and signalling each individual success with an individual burst of radiation.

It was found that there were three types of radioactive decay, each with a distinctive type of radiation associated with it. These were originally labelled alpha (α), beta (β) and gamma (γ). As one might guess, these rather bland labels were chosen to cover the original ignorance of the nature of the radiations, but the labels are convenient and are still used even though the phenomena are now well understood.

The radiation emitted in alpha decay is identical to the nucleus of the helium atom—that is, a composite particle of two protons and two neutrons. An alpha decay therefore transmutes the original isotope into one which is four atomic mass units lighter. Thus alpha decay of uranium-238 yields the 'daughter' isotope thorium-234, and alpha decay of radium-226 yields the 'daughter' isotope radon-222.

Beta radiation consists of electrons, either negatively charged electrons identical to those that occur in the electron clouds of atoms or the positively charged variety sometimes called positrons. Electrons do not exist as such in the atomic nucleus, so what is basically happening in beta decay is that one of the neutrons in the nucleus of the original or 'parent' isotope is transformed into a proton and an electron is emitted, or a proton is transformed into a neutron and a positron is emitted. A beta decay therefore transmutes the parent isotope into a daughter isotope of a different element but with the same mass. Thus potassium-40 beta decays to argon-40, and lead-212 beta decays to bismuth-212.

Gamma radiation is somewhat different in that it usually occurs in association with the other two. It is a burst of electromagnetic radiation similar in character to light or radio waves, but with a much shorter wavelength. It does not therefore result in the transmutation of the original isotope. After an alpha or beta decay, the nucleons in the daughter nucleus 'settle down' into their minimum energy state, emitting any excess energy as one or more gamma rays.

It is interesting to note that these three forms of radioactive decay involve three of the four different types of force known in nature. Alpha decay proceeds via the strong interaction, beta decay via the weak interaction and gamma decay via the electromagnetic interaction. Their detailed study can therefore give us, and has given, a deeper understanding of the basic forces that drive the Universe.

We have seen that alpha particles each carry two positive units of electrical charge and beta particles one unit of negative charge (for electrons) or positive charge (for positrons). Being electrically charged, they produce a phenomenon known as 'ionisation' when they pass through matter. Hence they are sometimes referred to collectively as ionising radiation. What happens is that because of their electrical charge they can disrupt the electron clouds of the atoms in the material they are passing through as they fly by. This disruption, or rather the rearrangement of the atom after the event, can manifest itself either as an electrical discharge or a flash of light. Thus it is literally possible to 'see' individual alpha or beta particles, small as they are, from the ionisation track they leave behind them. Electromagnetic radiation would not normally be classified as ionising because it does not carry an electrical charge, but because gamma radiation is such a high-energy form of electromagnetic radiation it can undergo processes which enable it to be regarded as such. When passing through matter, particularly heavy matter, it can eject an electron from an atomic electron cloud, or it can in some cases spontaneously convert itself into an electron–positron pair. The electron or electrons so produced are then capable of causing ionisation.

Because of its ionising effect, radiation is extremely easy to detect. Two basic types of detector have been developed: the Geiger counter, which works on electrical discharge, and the scintillation counter, which works on the detection of light flashes. The sensitivity of these detectors is phenomenal. You may have heard the characteristic 'clicks' of a Geiger counter. Each click represents *one* particle from the transformation by radioactive decay of the nucleus of *one* atom. To anyone more familiar with the detection techniques of traditional chemical analysis, the ability to detect a single atom might seem almost incredible. Strictly speaking, of course, by the time you have detected it by means of its decay particle it has by definition just ceased to exist, but that is a rather philosophical point.

The ionising effect also takes its toll on the particle itself, and rapidly robs it of its energy. Thus particles from radioactive decay can only travel limited distances through matter. Alpha particles, carrying twice the charge and about eight thousand times the mass of a beta particle, have by far the shortest range. Typically they can travel only a few inches in air, and can be stopped by a thin sheet of paper. Beta particles are able to travel a little further and are able to traverse a thin sheet of a light metal. Gamma rays have the longest 'range', though because of the second-order process whereby they lose their energy they do not have a precise range as such. It often takes an inch or so of a heavy metal such as lead to stop them. We shall be returning later to the ranges of these various types of radiation when we consider the steps taken to protect people from the injurious effects of high levels of radiation.

If one were to isolate a fixed amount of a given radioactive isotope, the amount would not remain 'fixed' indefinitely. By definition, if it is decaying, transmuting into another element, the amount of the original isotope, and consequently the rate at which it emits its radiation, must decrease with time. A single radioactive atom taken in isolation has a fixed probability of decaying in a given interval of time; this probability is independent of how long the atom has previously existed and of how many other atoms there are around it. It depends only on the type of atom it is. Consequently, for the large number of atoms necessary to make up a macroscopic piece of the substance, the rate of decay, and hence the rate of release of radiation, will be proportional to the number of atoms existing in the piece at that point in time. This gives rise to what is known as an exponential rate of decay, which can be characterised by a parameter called the 'half-life'. After one half-life, the quantity of the original radioactive substance will have reduced to one half, as also will the rate of release of radiation. After two half-lives, these will have reduced to a quarter, and after three half-lives to one eighth and so on.

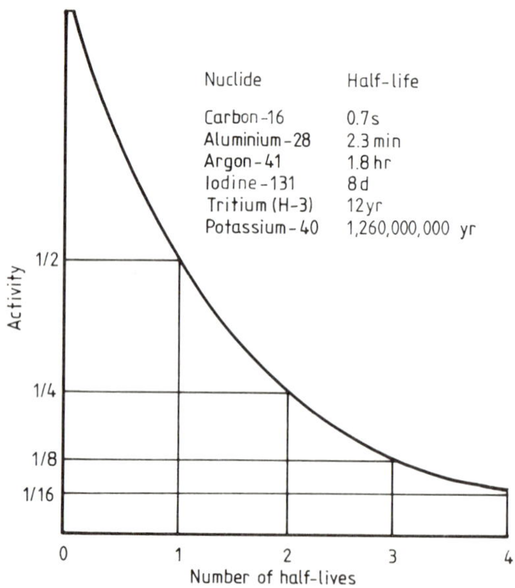

Each radioactive substance is characterised by its 'half-life', which is the time taken for half of its radioactivity to decay away. After two half-lives the radioactivity has fallen by a half times a half, i.e. a quarter, and so on.

The half-lives of the different radioactive isotopes vary enormously. Some, such as polonium-216, have half-lives of only a fraction of a second whereas some others, such as uranium-238 and uranium-235, have half-lives of four and a half billion years and seven hundred million years respectively. The shorter the half-life the faster the rate of decay and hence the greater the rate of emission of radiation. Thus a sample of iodine-131, with a half-life of only eight days, will be fiercely radioactive whereas the same quantity of natural uranium, which is predominantly uranium-238, will be only slightly radioactive. In the limit, of course, an isotope with an infinite half-life will not be radioactive at all. The significance of this inverse relationship between half-life and activity has a bearing on the management of radioactive nuclear waste, which we shall discuss later on.

We would only expect to find in nature those radioactive isotopes with half-lives that are at least a significant fraction of the age of the Earth, which is some four and a half billion years. Those with very much shorter half-lives would have decayed away a long time ago. Thus one would expect to find uranium-238, which has a half-life of about the age of the earth, and even some uranium-235, with its half-life of some seven hundred million years. However, one would not expect to find any iodine-131, with a half-life of eight days, and indeed one does not find it existing in nature. In practice, though, things are not quite as simple as that. For example, one might not expect radium-226, which has a half-life of only sixteen hundred years, to exist in nature. But, perhaps surprisingly, it does. How can that be? Why has it not all decayed away long ago? The answer is that it is being produced as fast as it decays. Like most naturally occurring radioactive substances it is part of a decay chain—a chain which begins with uranium-238, which decays to thorium-234, which decays to palladium-234, . . ., and so on down to lead-206, which is stable. As one link in this chain, thorium-230 decays to radium-226, and as the next link, radium-226 decays to radon-222. All the links in the chain achieved equilibrium long ago, and have thus settled down to abundances relative to each other that are determined by the relative half-lives. Thus, for example, radium-226 is produced at the same rate from radium-230 decay as that at which it disappears by decaying to radon-222.

Three such radioactive decay chains exist in nature. The

one we have just described, starting with uranium-238 and ending with lead-206, is known as the uranium–radium chain. The others are the thorium chain, which starts with thorium-232 and ends with lead-208, and the actinium chain, which starts with uranium-235 and ends with lead-207. As one moves down the chain, the changes in atomic mass are either zero or in units of four. This is because alpha decay involves the emission of an alpha particle which has a mass of four atomic units. Beta decay, because it involves an electron which has zero atomic mass, does not change the atomic mass of the decaying nucleus. This provides a convenient way of classifying the decay chains. Thus all the isotopes in the thorium chain are multiples of four; hence it can be labelled the '$4n$' series. The isotopes of the uranium–radium chain are multiples of four plus a constant value of 2; hence it can be labelled the '$4n + 2$' series. Similarly, the actinium chain can be labelled the '$4n + 3$' series. The careful reader will have noted an omission. What happened, one might ask, to the '$4n + 1$' series? The answer is that it must have existed once, but it all decayed away long ago. This series, known as the neptunium chain, starts with plutonium-241 and ends with bismuth-209. The isotope in the chain with the longest half-life is neptunium-237, but as this is only about two million years, it is not long enough to have enabled it to have survived from the formation of the Earth.

Because of the extreme sensitivity with which radiation can be detected, radioactive isotopes have a number of practical applications. Medical diagnosis is the most familiar. Radioactive isotopes have the same chemical properties as the stable isotopes of the same element. If they are administered to the body, even in minute quantities, they move through the system in the same way as the stable isotopes; this movement can be easily traced, and the performance of the organs involved can be deduced. One example is radioactive iodine, which is used to study the behaviour of the thyroid.

Similar 'tracer' techniques can be used to study phenomena in the natural environment, such as the movement of ground, water or river silt, and as a diagnostic tool in industry. As an example, let me describe a study performed at the naval base at Rosyth in Scotland. For many years it had been necessary to dredge the harbour to remove silt in order to keep it open for large ships. The spoils from this dredging operation were dumped at a point way out in the Firth of Forth. This dredging had to be done as a more-or-less continuous process, and the authorities were mystified as to where all the silt was coming from. So a small amount of a radioactive tracer—scandium-46—was added to one load of silt as it was dumped so that its subsequent movement could be investigated. With some consternation it was discovered that the combination of currents and tides was such that the silt drifted back from the dumping ground into the harbour. Apparently they had been shifting the same silt up and down stream for years! Using this technique it was also discovered that if they had chosen a different dumping ground some quarter of a mile or so from their traditional one, then the combination of currents and tides was such that the silt would have been swept out to sea. So one more dredging operation was performed, the spoils were dumped at the new spot, and as far as I am aware it has not been necessary to dredge the harbour again. However, shed a tear for the workers on the dredging machine who had previously taken home fat pay packets because of all the overtime but now find themselves without a job.

One naturally occurring radioactive isotope which has found useful application is carbon-14. This isotope has a half-life of some five thousand years or so, and would not exist at all in nature if it were not for the fact that it is continually being produced by the interaction of cosmic rays from outer space on the nitrogen atoms in the atmosphere. As the cosmic-ray flux is constant an equilibrium has been established, and hence all carbon in the natural environment contains a trace amount of carbon-14, about one part in a million million. All living matter continuously exchanges carbon with the environment, and so the amount of carbon-14 in such matter will be at the equilibrium level. When the organism dies this exchange ceases, and the amount of carbon-14 in the remains decays with a half-life of five thou-

Carbon-14 dating has made a valuable contribution to archaeological research. It isn't always as gruesome as this. (© UKAEA.)

sand years. By measuring the amount of carbon-14 in such remains, some estimate can therefore be made of its age. This is the basis of the carbon-dating technique, which has made such a valuable contribution to archaeological research.

The nuclear industry has to handle radioactive materials, and so it is necessary to have some knowledge of the basic principles of radioactivity in order to understand the techniques used. However, radioactivity as such has nothing to do with the basic process on which the generation of nuclear power depends. True, the isotopes of uranium are radioactive, but uranium-235 has a unique property that has made it possible for man to exploit the power of the atomic nucleus; this we will now describe.

NUCLEAR FISSION

The German scientist Otto Hahn had devoted much of his research career to the study of radioactivity. In 1938, while working with his colleague Fritz Strassmann at his laboratory in Berlin, he discovered a peculiar property of uranium-235. He found that under bombardment by slow neutrons the nucleus of this atom was able to split into two parts of roughly equal mass, with the emission, at least on an atomic scale, of prodigious amounts of energy, a process which was given the label 'nuclear fission'. It was later discovered that uranium-235 could undergo fission spontaneously, though not as readily as under neutron bombardment. At the time the discovery was regarded as little more than a scientific curiosity, and Hahn and Strassmann reported their findings in the open literature. In retrospect it seems somewhat bizarre that a discovery that was to lead to such dramatic consequences, and in view of the dense cloak of secrecy that was to descend on nuclear research, should have been announced to the world with such naive openness.

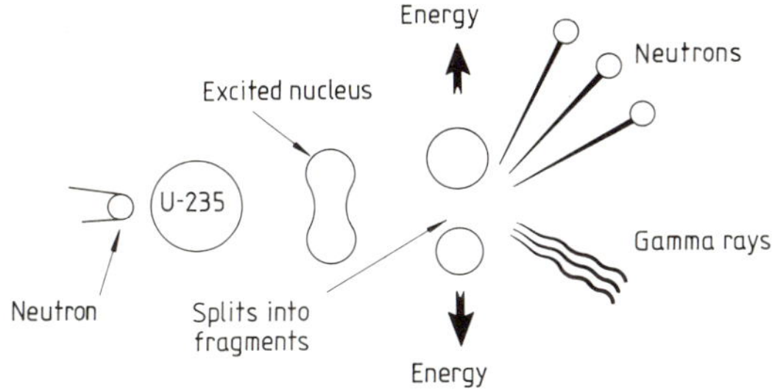

The fission process.

It did not take long for scientists around the world to appreciate the significance of the discovery. Apart from the fission fragments and the energy release, a few neutrons (two or three) are also emitted during fission of uranium-235. The neutrons can then strike other uranium-235 nuclei, causing them to undergo fission, leading to more energy release, and yet more neutrons which can then cause further fission and so on. So the possibility existed of establishing a self-sustaining chain reaction, rapidly multiplying the individual atomic-scale energy releases into a huge energy release on a macroscopic scale, the like of which the world had never seen.

One scientist who very quickly realised the significance and possible consequences of the discovery of fission was Lise Meitner. She had worked with Otto Hahn in Berlin, and was therefore very familiar with his work. She had been obliged to leave Germany in 1938 because of the rising tide of Nazi persecution, and was working in Copenhagen with her nephew, Otto Frisch. She reported her fears to the head of the Copenhagen institute in which she was working, the world-famous Danish physicist Niels Bohr. He too appreciated the significance of the discovery and confirmed her fears.

Lise Meitner. As one of the early pioneers of nuclear research, she quickly realised the significance of nuclear fission. (©L Meitner-Graf.)

The British scientific community was also fully aware of these developments, and of the eminence of Niels Bohr in the rapidly growing field of nuclear science. With the outbreak of the Second World War, and the subsequent German invasion of Denmark in 1940, an operation was mounted to extricate him from under the noses of the advancing Germans. The task was undertaken by the RAF in an incident which has produced its own contribution to the folklore of the early days of nuclear development. After Bohr had been picked up successfully, the plane had to fly at high altitude on its return flight to Britain in order to avoid interception—at such a height, in fact, that oxygen masks were necessary. It was then discovered with some horror that Bohr had an exceptionally large head, and there was no way it could be fitted into the standard-issue oxygen mask! There was nothing for it but to allow him to pass out through lack of oxygen, in the hope that he would come round again at ground level. Spare a thought for the predicament of the crew. There they were, speeding across the North Sea carrying a vital and invaluable cargo—one of the best nuclear brains in Europe—which was likely to expire at any moment, and all for the want of an outsize oxygen mask. Fortunately he recovered, and suffered no permanent ill effects from his experience.

From Britain Niels Bohr was able to contact Albert Einstein, another refugee from Nazi persecution, who at this time was in the USA. Einstein sent a now famous letter to President Roosevelt, urging him to give top priority to the development of the atomic bomb. Thus the Manhattan Project was born—the largest scientific collaboration in the history of mankind, with contributions from high-calibre scientists from Britain and refugees, like Bohr and Einstein, from occupied Europe and financed by the USA.

One can see why the Project was motivated by gut fear. Nuclear fission had been discovered by German scientists, and German science was as advanced as any in the world. It was inconceivable that the Germans would not attempt to harness this force and develop atomic weapons. The prospect of a Nazi Germany armed with atomic weapons was almost too horrible to contemplate. They were led by a madman who had at his disposal the most efficient military machine yet seen. They had taken on virtually the rest of the world and for a while it looked as if they might win. They had to be stopped. But how? The only way was for the Allies to get there first. Within the context of the knowledge available at the time the logic seemed impeccable. Fears were confirmed by intelligence reports that Germany was importing vast quantities of uranium, and by propaganda statements that a new 'secret weapon' which would turn the tide of the war, which by then was against them, was just around the corner.

With the collapse of Nazi Germany in May 1945, the Allies discovered to their surprise that the Germans were nowhere near developing an atomic bomb; the reasons for this have never been explained satisfactorily. Some German scientists have claimed that it was because they weren't really trying; the truth is probably much more fundamental. The basic science of nuclear fission had become world property, but the engineering problems to be overcome were daunting, requiring an effort on the scale of the Manhattan Project. With the benefit of hindsight it now seems almost ridiculous that anyone could have thought it conceivable that Nazi Germany could have mounted its own Manhattan Project without detection at a time when it was encircled by Allied armies and was being blasted continually by the Allied air forces.

The much vaunted 'secret weapon' turned out to be the V1 and V2 rockets. Had the war continued much longer, they could well have become a formidable threat—witness the dramatic post-war development of missile technology. The only development of nuclear energy that had taken place was an attempt to use it as a source of automotive power, to supplement the country's diminishing petrol stocks. The vast imports of uranium, as it is a heavy metal, were intended to make extra shielding for tanks. Perhaps they really weren't trying.

So what was to be done with the Manhattan Project? The German threat, against which it was supposed to be the defence, had not materialised. With all the momentum it had gathered, it could not be stopped. Anyway, the war was not over—there was still Japan to contend with, and after that who knew what kind of a world there would be? So the Project continued at full speed. At the Alamogordo Bombing Range in the New Mexico desert, at 04.00 on 16 July 1945, the first nuclear weapon was tested under the code name 'Trinity'. It had a traumatic effect on those who observed it.

The scientists knew it would make a big bang: they had even calculated just how big a bang, but cold equations cannot convey fully the reality of actually seeing it happen. After the initial joy of success, reaction set in. Was someone actually going to drop one of these things on real people? Led by no less a person than Albert Einstein, whose personal intervention had been largely instrumental in starting the project, representations were made to the President to implore him not to use the weapon in anger against Japan. It was a difficult decision to make, particularly for a new President. Harry S Truman had only just succeeded to the presidency following the untimely death of Roosevelt and now he was faced with this. How does one balance the undoubted death and destruction that dropping the bombs would cause against the estimated loss of life amongst Allied forces, of perhaps a million or so, that might result from the full-scale invasion of Japan which might be necessary if the bombs were not dropped? Truman was a man who had the nerve to make a decision and to shoulder the consequences. He decided to drop the bombs. The war ended within a week, and the world has lived under the shadow of the Mushroom Cloud since. It will continue to do so for the rest of its history, for despite good intentions, what has been discovered can never be undiscovered.

I have gone into some detail in describing the events surrounding the development of the first atomic bombs because later on we must discuss an issue that rightly concerns many people: the possible proliferation of such weapons, and the effect that the existence of a worldwide civil nuclear power programme might have on that process. From a study of the history of how bombs were first developed certain principles emerge which will guide us in our discussion. That discussion must also be guided by the natural world as it is and human nature as it is.

Let us return for a moment to the basic process of nuclear fission. Uranium-235 is the only naturally occurring isotope that can undergo the process to any significant extent, and is therefore the only naturally occurring nuclear fuel. There are only two other fissile isotopes of any significance, plutonium-239 and uranium-233, but these have to be made artifically in a nuclear reactor. The former plays an important role in the civil nuclear power programme, and holds the key to its future development. The latter could become important if an alternative fuel cycle based on thorium is eventually developed.

One does not always get the same fission products following each individual fission event, as there is only a probability that a specified fission product is produced. This is because the process is basically statistical in nature, and depends upon the internal dynamics of the uranium-235 nucleus which are exceedingly complex and at present only partially understood. Thus when a macroscopic quantity of uranium undergoes fission one obtains a wide spectrum of fission products containing a score or more individual isotopes. The shape of this spectrum has been studied extensively and is well documented. From this basically phenomenological study one knows with some precision how much of a specific fission-product isotope will be produced from the fission of a given quantity of uranium-235.

Fission products are almost invariably radioactive, with half-lives ranging from a fraction of a second to tens of years. The two most significant fission-product isotopes, in that they are produced with relatively high probabilities and have significantly long half-lives, are strontium-90 and caesium-137, both of which have half-lives of around thirty years. The fission products provide a vital link in the chain whereby the energy associated with the strong nuclear force is transformed into a form which can be used by man. Following the fission of a uranium nucleus, the two fission products fly off back-to-back, carrying with them the energy released in the process as their speed of motion—their kinetic energy. This speed of motion is rapidly dissipated as the fission products collide with other atoms in the neighbourhood, thereby causing these neighbours to increase their kinetic energy. This increase in the motion of the atoms of a substance is what we experience on a macroscopic scale as heat. The fission products, therefore, are instrumental in converting the energy released in nuclear fission into usable heat.

Because of the statistical nature of the fission process, the number of neutrons emitted varies from one fission event to the next. It is typically two or three, and the average for the fission of a macroscopic piece of uranium is about 2·5. As we saw earlier, these neutrons play a vital role in that they can induce other uranium nuclei to undergo fission and thereby establish a self-sustaining chain reaction. However, there are other fates that could befall these neutrons. Some can escape from the piece of uranium before they have had a chance to cause another fission, some can be absorbed by the non-fissile isotope of uranium or by structural material in the reactor, and indeed some can decay—the half-life of a free neutron is about ten minutes. 'Neutron economy'—the management of the neutrons from the fission process—is vitally important in the design and control of a practical nuclear reactor, as we shall see later.

BREEDING NUCLEAR FUEL

There is another reaction which takes place in fission reactors which we must consider. One of the possible fates of neutrons from the fission of uranium-235 was the capture of a neutron by the non-fissile isotope of uranium, uranium-238. This does not represent a total loss, because the capture process leads to the formation of a useful new element. Adding a neutron, which has one unit of mass, to uranium-238 gives us uranium-239. This decays rapidly by beta emission to a new element, number 93—neptunium. As beta decay does not change the atomic mass, the isotope of neptunium so produced also has mass 239. Neptunium-239 is radioactive and beta decays with a half-life of two days to yet another new element, number 94—plutonium. Again, the isotope with the same mass, 239, is produced.

The significance of this chain of events is that the isotope plutonium-239 is capable of undergoing fission, and so can be used in its own right as a nuclear fuel. We therefore have a mechanism whereby the non-fissile isotope uranium-238, which you will recall comprises 99·3% of natural uranium, can be converted into a fissile material, plutonium-239. We thus have the possibility of using the whole of natural uranium for the purposes of energy production. Needless to say, this has a dramatic effect on the energy supply situation, and we shall consider this further when we discuss the fast reactor, which is designed to take maximum advantage of this process.

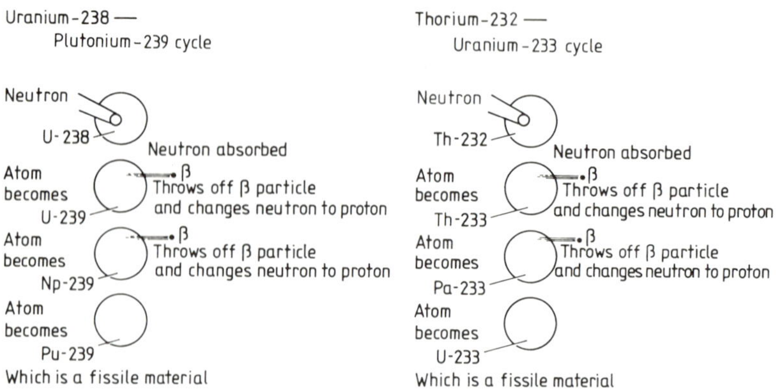

Plutonium-239 is only slightly radioactive. It is an alpha emitter with a half-life of 24 000 years. During the few years or so that a piece of uranium fuel stays in a reactor this isotope hardly decays at all, and so is available to capture a further neutron to form plutonium-240, and then another to form plutonium-241 and so on. These plutonium isotopes can beta decay to form yet more new elements, such as element number 95—americium—and element number 96—curium. These transuranic elements, that is those with atomic number greater than that of uranium (92), are referred to collectively as the 'actinides'.

When a piece of pure uranium is used as a nuclear fuel in a reactor, after a period of time it thus becomes contaminated with increasing amounts of two groups of other elements—the fission products from the fission of uranium-235 and the actinides arising from successive neutron capture in uranium-238. These two groups of elements form the nuclear waste arising from the use of uranium as a fuel (with the exception of the plutonium, which can be extracted and used as a fuel). The fission products, which form by far the larger portion, have half-lives of tens of years, whereas the actinides have half-lives of tens of thousands of years. We shall call on these basic facts later when we come to deal with the management of nuclear waste.

Isotopes such as uranium-238 from which fissile material can be bred, are referred to as 'fertile' isotopes. There is one other fertile isotope which exists in nature: thorium-232, which is the only isotope of natural thorium. Thorium-232 can capture a neutron and thereby become transmuted into uranium-233, which is an isotope capable of undergoing fission. To date, very little work has been done on this thorium cycle. Should it ever be developed fully, the world's thorium reserves would become available as an energy source, hence greatly enhancing the world's possible energy supplies.

HOW A REACTOR WORKS

Having seen how nuclear energy can in principle be produced through the process of uranium fission, we must now go on to see how this can be achieved in practice. The device through which this is done is called a nuclear reactor, though in the early days it was called an atomic pile. The first reactor, in which a self-sustaining chain reaction was achieved for the first time, was built by the world-famous Italian physicist Enrico Fermi in 1942 as part of the Manhattan Project. He built it, of all places, in a squash court underneath the stand of a baseball ground in Chicago. His historic reactor no longer exists (neither, for that matter, does the baseball ground). However, a modest plaque can be found on the spot to commemorate his achievement. The first reactor to feed power into a public electricity supply system was the Calder Hall station on the Windscale (now called Sellafield) site in Cumbria, which first came on stream in 1956. Calder Hall needs no plaque to commemorate it because it is still there and working well, and likely to continue to do so for many years to come.

An artist's impression of the scene in Chicago on 2 December 1942 when man first achieved a self-sustaining nuclear chain reaction. (© Chicago Tribune.)

Let us now look at the various components of a nuclear reactor and see what functions they fulfil. We start with the fuel itself. As uranium-235 is the only naturally occurring fissile isotope, this limits our choice of material for the reactor fuel. As we saw earlier, the fissile isotope 235 only occurs with an abundance of 0·7% in natural uranium. It is possible to establish a self-sustaining chain reaction using fuel rods of natural uranium, and indeed most of the earlier reactors did so. Most of the later designs, however, use 'enriched' uranium, in which a partial separation of the uranium isotopes has been achieved by a process we shall describe later. Enriched uranium for use as reactor fuel contains typically 2–3% of the fissile isotope 235, which makes the establishment of a chain reaction considerably easier. In passing it should be noted that weapons-grade uranium requires an enrichment of over 90%, and we shall return to this point later.

Fuel elements are therefore constructed as rods of either natural or enriched uranium, in either metallic or oxide form depending on the type of reactor. The fuel rods are 'canned' in some suitable material to prevent the fission products, some of which are gaseous, from seeping out into the coolant stream. The choice of canning material is governed by its properties with respect to neutrons and by its strength and corrosion resistance. It is essential that it does not absorb too many neutrons, as this would impede the performance of the reactor. In some designs an alloy of magnesium is used, in others an alloy of zirconium, and in yet others stainless steel.

Neutrons are emitted from the fission process at quite high speeds, and are in fact classified as 'fast neutrons'. They are better able to induce further fissions at much slower speeds. An important component of a reactor, therefore, is the 'moderator', whose role is to slow down the fission neutrons to speeds at which they can be more effective. The moderator is made of material containing light elements, and the neutrons literally bounce around among these light atoms, sharing their energy of motion with them on each collision. The neutrons are therefore reduced in speed very rapidly to a value characteristic of molecules at the temperature of the moderator. This process is called 'thermalisation', and the neutrons which have undergone it are referred to as 'thermal' neutrons. Reactors using a moderator to thermalise the fission neutrons are called thermal reactors, and all the existing commercial reactors fall into this category. Some reactors use carbon as their moderating material, in the form of graphite. Others use water, either the normal variety or 'heavy' water in which the hydrogen has been replaced by its heavier isotope, deuterium. The core of a reactor thus consists of a block or tank of moderator into which the uranium fuel rods are inserted.

There has to be some method of controlling a reactor, of switching it on and off and of varying the power output—the equivalent of turning up and down the flame in a furnace.

This is done by the use of control rods, made of a material with an extraordinary capacity for absorbing neutrons. The two best materials for this purpose are boron and cadmium. With the rods fully in, they mop up so many neutrons that the chain reaction cannot be maintained and the reactor stops working; when they are fully out, the reactor is working at full power; and at intermediate positions, any desired power output can be achieved.

We now have an operational and controllable nuclear reactor, but how do we get the power out so that it can be used? The energy from the fissioning of the uranium fuel appears as heat in the fuel rods, and this is extracted by passing a coolant over them which picks up the heat and transfers it to a boiler in which steam is raised. From then on the hardware is identical with that of a conventional coal- or oil-fired power station. The steam is passed through turbines in which the heat energy content of the steam is converted into mechanical energy of rotation. The turbines then turn electrical power generators, whose electrical output is coupled to the public electricity grid.

The coolant can be either a gas or a liquid. Some designs of reactor use carbon dioxide gas and others use water. In order to get a sufficiently high rate of heat transfer from the fuel rods, the coolant has to be pushed over them at a pressure which is many times higher than atmospheric pressure.

The radiation from the core of a working reactor is intense, and comprises escaping fission neutrons and gamma rays emitted from short-lived fission products. Fission products also emit alpha and beta particles, of course, but as these have a relatively short range they are reabsorbed in the core of the reactor. To protect the operators from this intense radiation, the core of the reactor is surrounded by a reinforced concrete shield typically twelve feet in thickness.

This then is in essence what a nuclear reactor is and how it works. There are, of course, a variety of permutations and combinations of fuel type, moderator and coolant in the commercial reactors currently in operation in various countries of the world. Often the detailed choice has been determined either by history or commercial considerations rather than by engineering performance, as we shall see when we look at different reactor types in the next chapter.

THE NATURAL REACTOR AT OKLO

Time and time again in the development of science it has been the observation of the small and apparently insignificant that has led to major discoveries: we saw earlier how Becquerel discovered radioactivity almost by accident. At about the same time, the 1890s, another apparently insignificant observation was made in the neighbouring discipline of chemistry. It was found that nitrogen gas obtained from the atmosphere was slightly heavier (about half a per cent or so) than that obtained from chemical reaction. In 1894 two English scientists, Lord Rayleigh and Sir William Ramsey, investigated this apparently insignificant problem and found that atmospheric nitrogen contained a small amount (approximately 1%) of a very much heavier and previously unknown gas—argon. One reason why it had not been discovered before was that it does not, to all intents and purposes, undergo any chemical reactions at all. It is completely inert, and is one of a series of inert gases called the noble gases, the other members of which (helium, xenon, neon, krypton) were then discovered rapidly. It is surprising how much easier it is to find things when you know what you are looking for. The discovery of these noble gases led to a much deeper understanding of the basic nature of atomic structure.

Ironically, argon should have been discovered about a century earlier. In 1785, another English scientist, Henry Cavendish, performed an experiment to try to find out whether the gas that was left after the oxygen had been removed from air was a single substance or a mixture of substances. He reacted the 'deoxygenated' air to remove nitrogen. Virtually all the gas disappeared, so he concluded that nitrogen was the only component of air other than oxygen. In fact he did have a tiny bubble of gas left at the end of his experiment, but because it was less than 1% of what he started with he ignored it! That tiny bubble of gas was, of course, argon: he had isolated the world's first sample of a noble gas, but had then ignored it because it was so small. One wonders how many other great discoveries were never made in history because of insufficient attention to detail. Winston Churchill is reputed to have said that man sometimes stumbles across the truth, but he usually picks himself up and walks on.

The charge of insufficient attention to detail could hardly be levelled at a French scientist, H Bouzigues, who was working at a laboratory at Pierrelatte in France in 1972. He was involved in the routine analysis of uranium samples, which amongst other things included a check on the ratio of the two isotopes of uranium, 235 and 238. This 235/238 ratio is almost a constant of nature, applicable not only to the Earth but also to the solar system as a whole. Its value is 0·7202% and it is known very accurately, the uncertainty being only six units in the fourth significant figure, which can be expressed in a convenient shorthand as $0{\cdot}7202 \pm 0{\cdot}0006\%$. It was found to be virtually the same for all uranium deposits that had been found on Earth. It is also the same for the traces of uranium found in the samples of lunar material brought back by the Apollo missions, and in similar traces found in meteorites.

However, Bouzigues found that the sample he was analysing one day in May 1972 had a 235/238 ratio of $0{\cdot}7171 \pm 0{\cdot}0007\%$. Was the discrepancy between this and what he expected significant? After all, it was only three parts in a thousand, and only just over twice the sum of his experimental uncertainty and the uncertainty on the expected value. No one would have blamed him if he had ignored it, and put it down to some unknown error in his measurement. But he did not ignore it, and a scrupulous investigation followed, leading to an exciting and almost incredible conclusion.

The sample which was analysed that day in May 1972 contained uranium coming from several mines, and so considerable scientific detective work was required to trace the anomalous uranium back to its mine of origin. This was eventually found to be a mine at Oklo in the West African state of Gabon. A test bore sunk at the mine revealed a 235/

238 ratio of only 0·44%. Other samples showed even lower ratios, the lowest recorded being 0·296%—a far cry from the supposedly 'constant' value of 0·7202%. So what started off as a puzzling small discrepancy in a laboratory in Pierrelatte had become a major phenomenon at Oklo—but still equally puzzling. A large fraction of the 235 isotope had disappeared from the uranium at Oklo. What could possibly have been the cause? The chemical and physical processes that go on in geological and climatological changes cannot be expected to discriminate between isotopes of the same element, particularly between isotopes so nearly equal in mass as those of uranium. What process had been at work differentially to remove uranium-235 and leave uranium-238 behind?

The solution of the problem seems so far-fetched as to be consigned to the realm of science fiction, were it not for the fact that the evidence for it is so overwhelming that none of the scientists investigating the problem have any doubt that it must be true. The ore at Oklo contains uranium in a very high concentration, about fifty times more so than one usually finds in bodies of uranium ore. It is not difficult to conceive of natural processes that could lead to such a high concentration. The ore contained none of the elements known to be very efficient at absorbing neutrons, but it did contain plenty of water in the crystal structure of the ore and water is known to be a good moderator of neutrons. Again, there is no reason why well understood natural processes should not have created these conditions also. So what must have happened at some time in geological history is that the body of ore went critical and started operating as a natural nuclear reactor, causing the uranium-235 in the ore to undergo fission!

Over the past hundred and fifty years or so, scientists have continually been digging up all sorts of interesting fossils, but the discovery of the fossilised remains of a nuclear reactor is surely enough to stretch the mind of even the most gullible. But there can be no doubt that is what has been found in the body of ore at Oklo. Subsequent research has filled in the details of this remarkable phenomenon. The Oklo reactor started operating about 1800 million years ago, and continued to do so intermittently over the next 700 000 years at an average thermal power output of about 20 kilowatts (kW). Some 800 tons or so of uranium were involved, thus fissioning about 6 tons of uranium-235, and something like a couple of tons of plutonium were produced. All this plutonium has, of course, long since decayed as its half-life is only 24 000 years, but the elements into which it decayed can be clearly seen, as also can the decay products of the fission fragments of the uranium-235. The volatile fragments such as iodine and the noble gases have escaped, but the others remained *in situ*. Neither they nor the plutonium had diffused more than a centimetre or so from the point at which they were formed. This latter discovery is of importance in the consideration of the geological disposal of nuclear waste, as we shall see later.

Exactly how had the random forces of nature managed to put together a working nuclear reactor when man-made reactors need such precision engineering? In order to get a reactor to work with natural uranium one needs either a heavy-water moderator or a very precise geometrical arrangement of the fissionable material. It is not credible that either of these conditions could have been met in nature, so how did the phenomenon happen? The answer is that natural uranium today is not what it was 1800 million years ago. The 235/238 ratio is only a constant at a fixed point in time, for the reason that the two isotopes have different half-lives. Uranium-238 has a half-life of about four and a half billion years, whereas uranium-235 has a half-life of only seven hundred million years. So the 235/238 ratio decreases slowly with time. At the time of the creation of the earth it was about 17%, and by the time of the Oklo phenomenon 1800 million years ago it was down to 3%. Now at 3% of the 235 isotope in uranium it is not necessary to have heavy water to enable fission to start—ordinary water will do—nor is it necessary to have a precise geometrical arrangement of the uranium. All one needs is a sufficiently high concentration of uranium in the ore, the absence of neutron-absorbing materials and the presence of water—just the conditions, in fact, that were found at Oklo.

The minimum value of the 235/238 ratio for such a natural

reactor to work is about 1%, and that value was passed about 400 million years ago. There is thus no possibility of finding a natural reactor still working today, but the chances for finding the fossilised remains of those that might have operated in the past must be quite high.

When Enrico Fermi got his reactor to work under the stand of that baseball ground in Chicago in 1942, he thought that he had built the first in the world. In the event it now appears that nature beat him to it by nearly two billion years.

THE FAST REACTOR

We must now leave our brief but interesting sojourn into nuclear pre-history and return to our present age, for there is one aspect of nuclear technology which we have not yet discussed but which is vital to its long-term application.

Only uranium-235 can be used as a nuclear fuel. In a reactor, some of the uranium-238 is converted into plutonium-239, some of which burns up *in situ* and the rest of which is extracted for future use as a nuclear fuel. If this bonus is included then, roughly speaking, about 1% of the world's uranium supplies can be considered as nuclear fuel. From a knowledge of how much uranium there is in the world which can be extracted, it soon becomes clear that 1% of it can only provide a short-term benefit to mankind. All this had been obvious from the early days of the nuclear industry, and so, right at the beginning, thought was being given to the design of reactor systems that could make more efficient use of uranium.

In a reactor there are two processes going on at the same time—the fission of the nuclear fuel, uranium-235 and the production of plutonium by neutron capture in uranium-238. The conditions in a thermal reactor are such that the former reaction goes more readily than the latter. So fuel, in the form of uranium-235, is burnt up more rapidly than fresh fuel, in the form of plutonium-239, is produced. Is it possible to design a reactor in which the fuel production process goes more readily than the fuel burning process? The answer is yes, and the design is called the fast breeder reactor. It leads to the perhaps bizarre situation that one ends up with more fuel than one started with. We are not getting something for nothing, of course; the extra fuel arises from the more efficient conversion of uranium-238.

The design for this more efficient conversion is achieved in the following way. Neutrons emerge from the fission process at high speed—hence the adjective 'fast'—and in thermal reactors they are slowed down by a moderator so that they are more efficient at producing further fissions. In a fast reactor the moderator is dispensed with and the neutrons are used at their full energy. In order to sustain the chain reaction with these fast neutrons a much higher concentration of fissile material is required in the fuel, and the fuel rods have to be more tightly packed. The concentration, or enrichment, of fissile material in the fuel of a thermal reactor need only be 2 or 3%, but in a fast reactor it must be as high as 20%. In practice this is achieved by constructing fuel rods for the fast reactor from plutonium, obtained from a thermal reactor, mixed with 'depleted' uranium, i.e. uranium from which most of the 235 isotope has been removed either by its previous sojourn in a thermal reactor or by an enrichment process to be described in the next chapter.

The 'core' of a fast reactor, then, consists of a compact assembly of fuel rods made up from a 20/80 mixture of plutonium and uranium, usually in oxide form. This is surrounded by a 'blanket' of depleted uranium in which neutrons escaping from the core are captured, leading to the conversion of some of this uranium to plutonium.

Because of the compactness of the core, the removal of the heat from it for electricity production creates a special problem. Water cannot be used as it is a moderator. One could use gas, but very high pressures would be required, and although it is possible in principle to achieve these, it is more usual to use a liquid metal, such as sodium, as the heat-transfer medium. At first glance the choice of sodium seems surprising—it does, after all, interact violently with both air and water—but it has many properties which make it a good choice. It has a high boiling point, and when pure it is in fact less corrosive than high-temperature steam. It must be well contained, of course, but as it is used at atmospheric pressure the problems in doing so are much less severe than for a high-pressure gas.

The sodium becomes radioactive. For this reason, and to prevent sodium/water reaction products arising from small leaks getting into the reactor core, the heat is transferred through a heat exchanger to a secondary sodium circuit, thence through another heat exchanger to a steam raising plant. The steam is then used to produce electricity as in any other power station.

Let us digress for a moment on the plutonium balance in a fast reactor system. It is sometimes called the 'breeder reactor' because it is capable of producing more fuel than it burns, and this has led some untutored minds to conjure up a situation akin to that of the Sorcerer's Apprentice and his problems with breeding broomsticks. The core of a fast reactor is a net consumer of plutonium and the blanket is a net producer. By appropriate design of the blanket, therefore, the amount of plutonium produced can be less than, the same as, or more than the amount consumed. During the stage in which one is building up a programme of fast reactors, an excess is required in order to provide the initial fuel charge for new reactors coming on stream. Most of this excess will be provided by thermal reactors because their net output of plutonium is greater (they produce it, but do not use it as their initial fuel). Eventually one reaches an equilibrium stage, at which a fixed amount of plutonium circulates round the cycle, and what is consumed is the major isotope of uranium, uranium-238. Finally, several centuries in the future when the world's supply of uranium becomes exhausted, the final charges of plutonium will be incinerated in 'blanketless' fast reactors.

Thus we see that with a fast reactor system we can make optimum use of the world's uranium resources. In principle, all the energy content of the uranium could eventually be used, but in practice it is more like 50% that can be obtained.

This is to be compared with about 1% for a thermal reactor system. In a world likely to be hard-pressed for energy, any system which can yield a fifty-fold increase in the utilisation of a primary fuel is one that we cannot lightly ignore.

NUCLEAR FUSION

We saw earlier that the natural elements occur in a series from the lightest (hydrogen, atomic number 1) to the heaviest (uranium, atomic number 92). If one looks at the nuclei of these elements one finds that those towards the middle of the series are more tightly bound than those at either the heavy or the light end. Indeed, this is the basic reason why energy is released in the fission of a uranium-235 nucleus. It splits into two fission products, which are middle-order elements with nuclei which are more tightly bound, and the excess energy is released. One is therefore led to ask if it is also possible to obtain an energy release by fusing together two of the lighter elements into a heavier one. The answer is yes, and the process is called nuclear fusion. Thus two of the isotopes of hydrogen, deuterium and tritium—hydrogen-2 and hydrogen-3 respectively—can fuse together to form helium-4 plus a free neutron, with the release of a considerable amount of energy. The process is extraordinarily difficult to achieve on a macroscopic scale. This is because of the sheer dynamical problems of getting the two nuclei together so that they can fuse, and also because one does not have the equivalent of the chain reaction which one has in fission to help to sustain the process. To date, nuclear fusion has only been achieved in an uncontrolled manner, and this is the basis of the hydrogen bomb.

In order to achieve controlled nuclear fusion it is necessary to compress the deuterium and tritium nuclei into a high-temperature plasma—a state in which the two atoms are fully ionised—and hold them there at a minimum density and temperature for a minimum period of time (known in the trade as the 'Lawson criterion'). Early in 1957 we thought that we had done it. This was the highly acclaimed Zeta project at Harwell. Free neutrons were observed escaping from the equipment—a sure sign, it was thought, that fusion had been achieved. Amidst the general euphoria that surrounds a scientific breakthrough, some pessimist asked if we were really sure that the neutrons we were seeing actually came from the fusion process, and not from some other effect. So the neutrons were studied in more detail, and it soon became obvious that they were most definitely not coming from fusion. It is now generally agreed that achieving controlled nuclear fusion is going to involve a long hard haul.

Even after the laboratory demonstration of controlled nuclear fusion, the engineering problems to be overcome to extract the energy are daunting. Most of the energy from the fusion process is carried away by the neutron, and so its energy of motion has to be converted into a usable form. There is also another task that the fusion neutron must perform: to replenish the fuel supply of the reactor.

Of the two fuels needed for the process, deuterium is plentiful as it occurs in sea water. Tritium, however, is radioactive with a half-life of about twelve years, and so it does not occur in nature (apart from miniscule amounts produced in the upper atmosphere by cosmic rays). It can be manufactured, however, by bombarding lithium with neutrons. So a practical fusion reactor must have some form of 'lithium blanket' surrounding the reaction zone, in which the neutrons from the fusion reactor can interact to produce tritium to be fed back in as a fuel.

Fusion research has progressed slowly but steadily over the past twenty-five years or so, but no one is under any illusion as to the enormity of the task. Fusion reactions drive the processes from which the Sun and other stars derive their prodigious energy. This I suppose provides some small crumb of comfort and encouragement—we know that in principle it can be done. However, no one seriously expects fusion to provide a practical source of power until well into the next century.

CHAPTER 3

The Nuclear Industry in Profile

In reality the nuclear industry is a fiction. We do not have a nuclear industry in Britain in the same sense that we have a steel industry or a coal industry. What we do have is a number of organisations, some statutory and some commercial, with an interest in the development and exploitation of civil nuclear power. In most cases these organisations have many other interests, and in some cases their nuclear involvement is but a small part of their total operation. It might be useful at this stage to list these organisations and to describe the nature and extent of their involvement in the nuclear power programme.

The United Kingdom Atomic Energy Authority (UKAEA) is a research and development organisation set up by Act of Parliament in 1954 to take over the atomic energy project which from 1946 to 1954 had been under the aegis of the Ministry of Supply. It is charged with the duty of conducting research into all aspects of the peaceful uses of atomic energy, to provide expert advice on such matters to government and to provide authoritative information to the public. Its major research establishment—one of the largest and most diverse scientific research laboratories in the world—is the Atomic Energy Research Establishment (AERE) at Harwell in Oxfordshire. It also has other important research establishments at Risley near Warrington, at Winfrith in Dorset and at Dounreay on the northern coast of Scotland; it has its headquarters in London.

British Nuclear Fuels Limited (BNFL) is an organisation—which split off from the UKAEA in 1971—whose role is to supply fuel services to the nuclear power programme, both in Britain and also to some extent to those in other countries as well. These include uranium enrichment, fuel fabrication, reprocessing of spent fuel and waste management. It has three major factories—Capenhurst in Cheshire which handles the enrichment, Springfields in Lancashire which fabricates fuel and Sellafield (formerly Windscale) in Cumbria which handles reprocessing and waste management. Its headquarters are at Risley alongside the UKAEA establishment. BNFL was set up and operates as a commercial company, although all the shares are held by the government. For the purpose of conducting its enrichment business, BNFL is part of a European consortium, the Uranium Enrichment Company (URENCO), in which its partners are West German and Dutch companies. URENCO has its British headquarters at Marlow in Buckinghamshire.

The National Nuclear Corporation (NNC) is responsible for the design, construction and commissioning of nuclear power stations. It is a consortium of a number of large engineering companies, with the UKAEA holding about one third of the shares on behalf of the government. It has offices at Risley, at Whetstone in Leicestershire and at Knutsford in Cheshire.

The generating boards are responsible for the ordering and operating of nuclear power stations to enable them to meet their statutory obligation to supply electricity to the National Grid as cheaply as possible. There are four such boards serving different areas of the country. The Central Electricity Generating Board (CEGB) is responsible for England and Wales. The South of Scotland Electricity Board (SSEB) covers the southern part of Scotland and the North of Scotland Hydroelectric Board (NSHEB) covers the northern part of Scotland. Most of the operating nuclear power stations in Britain are owned by CEGB or SSEB. These Boards also purchase electricity from large prototype reactors owned and operated by other bodies. Thus the CEGB buys electricity from the 200 megawatt (MW) Calder Hall Magnox reactor at Sellafield, which is owned and operated by BNFL, and from the 100 MW steam-generating heavy-water reactor (SGHWR) at Winfrith, which is owned and operated by the UKAEA. The SSEB buys electricity from the BNFL reactor at Chapelcross—a twin to Calder Hall—and the NSHEB from the UKAEA's 250 MW prototype fast reactor at Dounreay. In Northern Ireland the responsibility for the generation and supply of electricity rests with the Northern Ireland Electricity Service. They do not have any nuclear power stations in their system at present.

There are also a number of statutory bodies which exercise a regulatory role with respect to the development and deployment of nuclear power. They are all independent of the organisations listed above. Indeed, they are each ultimately responsible to a different government minister. On the production side of nuclear power generation, responsibility rests with the Secretary of State for Energy. On the regulatory side, it rests with a variety of other ministers.

The Nuclear Installations Inspectorate (NII) is responsible for issuing licences for the construction of nuclear power stations or ancillary nuclear facilities such as, for example, reprocessing or fuel fabrication plants. The licensing procedure is a continuous one. First the design has to be approved, then inspectors monitor each stage of the construction and conduct regular checks during operation. Unless the NII is satisfied, the plant cannot be built. It also has the power to close down any plant if it considers that the conditions of the operating licence are not being met. The NII is a branch of the Health and Safety Executive (HSE), which is ultimately responsible to the Secretary of State for Employment.

The National Radiological Protection Board (NRPB) and, to a lesser extent, the Medical Research Council (MRC) have the job of recommending radiation exposure standards based on internationally agreed criteria, which are then enshrined in Acts of Parliament. They attempt to ensure that no worker or member of the public will receive a dose above the legal limit, that no one will receive any radiation at all unless it can be shown that there is some compensating benefit, and that all such doses will be kept as low as reasonably achievable—known in the trade as the 'ALARA' principle. The NRPB is responsible to one of the Health Ministers, and the MRC is responsible to the Department of Health and Social Security.

The Ministry of Transport is responsible for enforcing the regulations controlling the transport of radioactive material and the Ministry of Agriculture, Fisheries and Food regularly monitors the effluent from nuclear establishments, their findings being published in an annual report. The Department of the Environment (or the Welsh and Scottish Offices in Wales and Scotland) plays a key role in authorising waste disposal.

It might seem that the nuclear industry is the most extensively regulated industry in the country. Arguably it is so, and the reason for this is largely historical. The potential dangers in a nuclear power programme were obvious from the outset; hence legislation to control them was brought in at an early stage. In contrast in many other industries developed under commercial pressures the dangers only became apparent when the industry was at full steam; hence the necessary health and safety legislation had to be introduced piecemeal and in retrospect—always an unsatisfactory procedure and one likely to be compromised by established commercial interests.

These then are the organisations which have an interest in the production or the regulatory side of the nuclear power programme, which we refer to loosely as 'the nuclear industry'. Clearly the regulatory authorities, or their parent bodies, have many interests other than nuclear power. To the generating boards nuclear power is just one of several methods at their disposal to produce electricity. Their preference for it is on the grounds of cost—in their judgment they can produce cheaper electricity from nuclear power stations than from other options (apart from hydroelectric power, the scope for further development of which is rather limited in the UK). The engineering firms which make up the NNC consortium can, of course, build things other than nuclear power stations, which is perhaps just as well as it is only recently they have received any orders for nuclear power stations. Even the UKAEA has other interests. For example, about half the work now being done at Harwell is non-nuclear work performed under contract to other government departments or to private firms.

Although the nuclear industry is a fiction, it is a very convenient one and we can use it as a framework to describe the steps involved in providing useful energy from the nuclear fission process. The nuclear industry operates on a complete fuel cycle. The raw material is extracted from the earth and is fabricated into fuel suitable for burning in nuclear reactors. After several years in the reactor the fuel rods are removed and reprocessed; this involves the chemical separation of the unburnt fuel, which can then be recycled, from the waste products. The waste is initially stored in repositories on the surface and may eventually be disposed of either deep underground or at sea. Thus the business starts with the extraction of raw material from the earth and ends with the disposal of the waste material back into the earth. Let us now look at the steps involved in the nuclear fuel cycle in a little more detail.

URANIUM MINING

Uranium is the raw material of the nuclear industry. It is present in the Earth's crust at an average abundance of a few parts per million. That means that there are about 30 million tons of uranium in the earth. Most of it, of course, occurs at very low concentration and its extraction would be difficult and expensive. High-grade ores, with a concentration of 1000 parts per million or greater, occur in a number of places on Earth. Known reserves of uranium in these high-grade ores are some few million tons. Uranium prospecting is still in its infancy, however, and on geological evidence it is expected that the total amount of extractable uranium occurring in high-grade ores will be several tens of millions of tons.

How does that compare with the demand for uranium to fuel the nuclear power programmes of the world? In Britain we use about three thousand tons a year. The Western World as a whole—in official parlance, the World Outside the Communist Area (WOCA)—consumes about forty thousand tons a year. Little is known in detail about consumption in the Communist countries. Their officialdom is even less forthcoming than ours when it comes to such matters, but for the purposes of our present comparison it is reasonable to assume that their demand will be roughly similar to that of WOCA. So we see that there should be enough uranium in high-grade ores to supply the entire nuclear power programme at *ten times* its present level for something like fifty years. That represents an energy resource of formidable proportions, and is roughly comparable with the remaining reserves of oil. If that uranium were to be used in fast reactors rather than our present thermal reactors, its availability would be multiplied by a factor of 50 or 60. That pushes it way beyond the planning horizon of even the most far-sighted pundit.

Uranium ore. This is a piece of pitchblende from the Congo. (© UKAEA.)

The bulk of the high-grade ores in WOCA are found in the USA, Canada, Australia, South Africa and Namibia. From a purely nationalistic point of view, it will be noted that most of these countries have a long history of friendly relations with Britain—certainly more so than the countries that hold the bulk of the world's residual oil reserves. Of the European countries, France has significant indigenous supplies of uranium, and has access to supplies in former French colonies in West Africa, and there are useful deposits in Sweden.

Uranium occurs in nature in a wide variety of mineral forms, but they can be roughly grouped into two categories. First, there are those formed in an environment of rich organic matter. Perhaps the best known of this type is pitchblende, which contains a mixture of oxides of uranium. Other examples are coffinite, which is uranium silicate, and brannerite (found extensively in the Canadian deposits), which is uranium titanate. Uranium in this type of mineral is also found in association with coal and bituminous hydrocarbons in some of the deposits in the USA. The second category comprises minerals that have been exposed to extensive weathering and are hence in a high state of oxidation. The most important of these are the phosphates, such as autunite (uranium and calcium) and chalcolite (uranium and copper).

When prospecting for uranium, there is one particular property of these ores which is of great assistance—they are radioactive. Radiation can be detected with high sensitivity using very simple instruments and hence a radiometric survey of a likely area is usually the first step taken to locate a uranium ore body. As we saw in the previous chapter, uranium in nature exists in radioactive equilibrium with a

Prospecting for uranium. Here radon gas, one of the decay products of uranium, is being detected. (© UKAEA.)

chain of daughter products, all of which (with the exception of the end of the chain) are radioactive. One such daughter product is the gas radon. Being a gas, it can diffuse through the overburden of soil to the surface, where it can readily be detected. Radon acts as a convenient radioactive tracer and can be of great help in the search for uranium deposits in nature.

Once a uranium ore body is located, the next step is to 'prove' the deposit, i.e. determine its size, depth and quality in order to form a judgment as to whether or not it is workable. One also needs to know the characteristics of the ore to determine the best method of mining it and how it will be physically and chemically processed once it is raised to the surface. This is done by sinking a systematic pattern of boreholes from which core samples can be extracted for chemical analysis, using techniques rather similar to those used to prove deposits of other minerals.

Open cast uranium mining. This is the Mary Kathleen mine in Australia. (© Australian Information Service, London.)

Because of the wide variety of uranium ores existing in nature, the methods used to mine them vary. Sometimes the deposits are on or near the surface, in which case open-cast mining techniques can be used. In other cases, where the ore is deep down, mine shafts have to be sunk, often up to several hundred feet or, in some cases in South Africa where uranium can occur in association with gold, several miles.

An oxide of uranium called 'yellowcake'. It is in this form that uranium is transported from the mine to the user. (© UKAEA.)

As the concentration of uranium in the ore is usually only a few parts per thousand, the vast bulk of the material extracted is spoil. It is necessary to concentrate the uranium in mills built adjacent to the mine before it is transported to the user. First the ore is passed through a radiometric lump-sorting grader in which individual lumps of ore are selected or rejected on the level of their radioactivity. For some ores, those which have a much higher density than their host rock, a further concentration can be achieved on the basis of weight, but this method is not generally applicable. The selected ore is then ground up and dissolved in a solvent and the uranium concentrate extracted by a complex chemical precipitation and filtration process. The uranium ends up as an oxide—U_3O_8. This is usually in the form of a yellow powder known as 'yellowcake'. It is in this form that uranium is transported in drums to the user. It is only mildly radioactive, so no special precautions are needed during its transportation. There was one occasion when a barrel fell off its lorry whilst being transported and spilled its contents over the road. It was rapidly and safely removed by the traditional method of brush, shovel and bucket.

There are hazards associated with uranium mining. As with all mining operations, there is the risk of mine collapse or other physical accidents. It does not carry the risk of fire or explosion, which is an ever-present risk in the mining of coal, nor pneumoconiosis, another occupational hazard of coal mining. However, uranium miners, in common with others involved in hard-rock mining operations such as those to obtain tin, are subjected to the risk of lung cancer arising from the presence of radon. In the early days of uranium mining, the practices adopted left much to be desired, and a number of miners suffered from lung cancer as a result. Over the years, mining techniques have improved considerably, particularly mine ventilation, leading to a marked improvement in working conditions for uranium miners and a lower risk to their health. Improvements have also come about in the working conditions of coal miners over this period, of course. At the present time an uranium miner is at a level of risk of about ten times less than that of a coal miner, for the same energy content of fuel produced, i.e. about the same level of risk per ton of material mined.

Another hazard which can arise from uranium mining is that associated with the disposal of the spoil from the mine and the 'tailings', the waste product from the mills located at the mine head in which the ore is converted to yellowcake. This waste is mildly radioactive as it contains the products of the uranium decay chain and so some care has to be taken over its disposal. Again, in the early days insufficient care was taken, leading to justifiable concern from local residents over the effects that this might have on their health. Present-day practices are much improved. As with many problems in life, once the problem has been appreciated remedial action can be taken.

So far we have considered only the availability of high-grade ores—those with concentrations in excess of 1000 parts per million. There is, of course, no physical discontinuity in the concentration spectrum at this level. The practical limit is determined by economics. It is only economically viable to mine ores down to this concentration at the present world price of uranium, which is about $40 per pound. Should the market price increase as the nuclear programmes in various countries expand, then it would become economic to mine poorer ores. In particular, should fast reactors be eventually used to any large extent, for which the raw price of uranium has only a trivial effect on the cost of the electricity they produce, then a large price rise could be tolerated, leading to a vast increase in the amount of uranium available. The concentration spectrum of uranium ores is such that a ten-fold reduction in the uranium concentration of economically minable ore leads to a 300-fold increase in the total amount of uranium available.

Another, as yet untapped, low-grade source of uranium is seawater. This contains uranium at very low concentration—about three parts per thousand million—but there is an awful lot of ocean: in the oceans of the world there is something like ten thousand million tons of uranium. The problem will be in extracting it. In principle the technique is possible but in practice the problems are daunting, due mainly to the vast amounts of water that would have to be processed. Needless to say, it would be expensive, probably about a hundred times more expensive than extracting uranium from high-grade ore. Very little practical work has been done on this technology, although the Japanese are building a small prototype plant.

It is interesting to note how nuclear programmes in different countries have been influenced by their access to uranium

A pilot plant for extracting uranium from sea water. The process is not yet economic on a commercial scale. (© UKAEA.)

supplies. The USA, which is sitting on about half of the Western World's uranium supplies, has not felt the need to press ahead with spent fuel reprocessing—they simply store their spent fuel rods—nor with commercial application of the fast reactor. Similarly, in Canada, which has vast reserves of uranium in relation to its population, spent fuel rods are stored and there is little or no interest in fast reactor development. Also, as they did not have access in the early days to enrichment technology, they pioneered a reactor system using natural uranium, known as CANDU. In Britain and France who have respectively negligible or modest indigenous supplies of uranium, reprocessing technology is well advanced, as is also the development of the fast reactor. Japan, which has virtually no natural resources at all, is keen to lay its hands on any energy source it can. The Japanese have been willing to sign lucrative contracts with Britain and France for reprocessing work, as they have not yet fully developed this technology themselves, and they are also, as we saw earlier, investigating the possibility of extracting uranium from sea water.

In conclusion there is one final source of high-grade fissile material which we must mention. That is the plutonium and highly enriched uranium locked up in the nuclear warheads in the military arsenals of the world. It has been calculated that the Russians and the Americans each have enough fissile material tied up in their nuclear weapons to supply all of the electricity requirements of both countries for about ten years. Such is the price we are obliged to pay in resource terms because of suspicion and mutual mistrust.

A piece of uranium metal. (© UKAEA.)

ISOTOPE ENRICHMENT

It is possible to build a reactor using natural uranium, containing only 0·7% uranium-235, as its fuel, and indeed many of the early types—the Magnox in Britain and the CANDU in

Canada—did so and continue to do so quite successfully. However, there are many advantages in using uranium fuel with a higher proportion of uranium-235. For example, it enables the reactor core size to be reduced with a consequent reduction of the capital cost of the core vessel and its containment. It allows a greater 'burn up' before the fuel needs to be replaced and hence more energy can be obtained from each ton of fuel fed into the reactor. A higher proportion of fissile material means that it is easier to maintain the chain reaction, and performance is less sensitive to the geometric arrangement of the core and to the level of impurity in the various components. Therefore most of the later reactors, and those planned or under construction, use uranium fuel with a higher proportion of the 235 isotope than that which occurs in natural uranium; this is known as 'enriched' uranium. Typically they use uranium enriched to about 2 or 3%.

How does one achieve enrichment? Uranium-235 and 238 are isotopes of the same chemical element, and thus have identical chemical properties. One therefore has to devise some physical means of exploiting the small mass difference between the atoms of the two isotopes. As that mass difference is very small—somewhere around 1%—one can see that the task is non-trivial. Nevertheless it is possible in principle and has been achieved in practice.

The first method used was that of electromagnetic separation. When uranium atoms are ionised (that is, one of their outer electrons is stripped away, leaving it with a net positive electrical charge of one unit) they will be deflected when fired through a magnetic or electric field. By firing them through a correctly selected combination of electric and magnetic fields one can arrange that all the atoms of the same mass end up at the same place. One has built what is called a 'mass spectrometer', which sorts out a spectrum of atoms fed into it by depositing them at different places on the basis of their atomic masses. Electromagnetic separators have been used for many years (and are still used) to produce microscopic quantities of relatively pure samples of isotopes of various elements for research purposes. To do so on a macroscopic scale, as is necessary to produce usable quantities, say kilograms, of uranium-235, is a difficult engineering task, but it has been done and indeed some of the uranium-235 for the Manhattan Project was produced in this way. This technique has now been largely superseded by others.

The bulk of the enriched uranium for today's nuclear reactors is produced by the gaseous diffusion process. This works on the principle that gases diffuse through a porous membrane at different rates depending on their atomic or molecular weights, with the lighter gases diffusing at a faster rate than the heavier ones. Thus with a gaseous compound of uranium, the molecules containing the 235 atoms diffuse through a membrane faster than those containing the 238 atoms and hence a partial separation can be achieved.

A gaseous compound of uranium suitable for this purpose is uranium hexafluoride ('hex'), a compound containing one atom of uranium and six atoms of the very light element fluorine. It can be manufactured from yellowcake by a series of standard, if rather complex, chemical reactions. Hex is actually a solid at room temperature, which makes it easier to handle on its way to and from the diffusion plant. It sublimes, that is it converts directly from a solid to a gas without going through the liquid phase, a property it has in common with a few other substances, e.g. carbon dioxide, at 56·5°C. It can thus be converted to a gas using standard industrial steam-raising plant, then reconverted back to a solid after enrichment by conventional industrial refrigeration equipment. As fluorine is such a light element compared with uranium, it does not seriously reduce the already small percentage difference between the molecular weight of the 235 and 238 compounds below that of elemental uranium.

The physical properties of hex therefore make it well suited for the diffusion process, but its chemical properties did initially cause many headaches. It is stable in dry air, but reacts vigorously with water. It also attacks most metals and common materials of construction. Special materials therefore had to be developed for the construction of the plant, and some alloys of nickel, copper or aluminium were found to be suitable. The problems posed by the chemical properties of hex, though difficult, have not proved to be insuperable.

Because of the small mass difference between uranium-235 and uranium-238, only partial separation can be achieved by a single stage of gaseous diffusion. In order to achieve the required enrichment, one therefore has to link together a large number of units in series in a cascade. From each stage, the gas which has been slightly enriched with 235 isotope is passed up the cascade to the next stage, and the fraction which has been slightly depleted in 235 is passed down. The stage separation factor is very small, only 1·004 (i.e. the gas passing up from one stage to the next is only 1·004 times more enriched than the entrant gas). To achieve an enrichment of, say, 3%, starting with natural uranium at 0·7%, one needs a cascade of more than a thousand stages. That means that a gaseous diffusion enrichment facility is a very large piece of industrial plant. In fact, one's first impression when visiting such a plant, say the one at Capenhurst in Cheshire where the

Isotope enrichment. This is the gas diffusion plant at Capenhurst in Cheshire. (© UKAEA.)

enrichment work in Britain is carried out by BNFL, is its overwhelming size.

Isotope enrichment. This is the gas centrifuge plant at Capenhurst in Cheshire. (© BNFL.)

Another enrichment technique which has been developed and is now in commercial operation is the gas centrifuge process. The hex is spun round at high speed in a cylinder where, because of the centrifugal force, the heavier molecules containing uranium-238 tend to collect towards the outside of the cylinder and the lighter ones containing uranium-235 tend to collect near the centre; hence a partial separation is achieved. Again the individual cylinders are arranged in a cascade; the gas slightly enriched at the centre of each cylinder is passed up to the next stage, and that slightly depleted at the walls is passed back to the next stage down. The stage separation factor is very much larger than in the diffusion process, being about 1·2. This means that in order to produce 3% enriched uranium from natural uranium only about 25 stages are needed rather than over a thousand. So a gas centrifuge plant is very much smaller than a gas diffusion plant, and can be fitted into a room—albeit a very large one—rather than covering several acres as is necessary for a diffusion plant. Furthermore, as the individual centrifuges can be mass produced, a centrifuge plant can be built in about half the time required to build a diffusion plant. A further advantage is that a centrifuge plant requires only about a tenth of the electrical power to drive it compared with a diffusion plant of the same output. In both cases, of course, the power requirement is trivial compared with that generated by a reactor using the enriched fuel. A reactor can generate enough power to provide itself with a year's supply of centrifuge-enriched fuel in about 25 hours.

The many advantages of the centrifuge over the diffusion process would seem to indicate that it must eventually become the preferred method for producing enriched fuel for the nuclear power stations of the world. It has been pioneered in Britain by BNFL in collaboration with Dutch and German firms through the umbrella organisation URENCO. Two plants are in operation, one at Capenhurst in Cheshire and the other at Almelo in the Netherlands. With this technological lead it is likely that Britain and her European partners will capture an increasing share of the world enrichment market.

There are two other enrichment techniques that are being researched at the moment. One is the fixed-wall centrifuge, sometimes known as the 'jet' or 'nozzle' process. Here the curved wall is kept stationary and the gas is blown past it at high speed. Again the centrifugal force tends to keep the heavier portion of the gas closer to the wall than the lighter fraction, and hence a partial separation is achieved. A process of this type is being pioneered in South Africa, but very few details have been released. The other experimental technique makes use of lasers. The frequency of the laser beam is tuned so that it is absorbed by one uranium isotope but not the other. This can lead to the preferential ionisation of one isotope, which can then be separated from the other by an electrical field. Both of these techniques are still very much at the laboratory stage, and it is as yet too early to judge their likelihood of success.

Depleted uranium, from which most of the fissile 235 isotope has been extracted, is a by-product of the enrichment process. This shows Britain's stock of depleted uranium, stored at Capenhurst in Cheshire. If used in fast reactors the depleted uranium shown is equivalent in energy terms to the whole of our recoverable coal reserves. Without fast reactors it is of little further use. (© BNFL.)

REACTOR FUEL FABRICATION

The production of reactor fuel elements in Britain is carried out at the Springfields Works in Lancashire, a plant operated by BNFL.

The process starts from uranium ore concentrate (yellow-cake), which is the form in which it is delivered from the uranium mines in various parts of the world. The concentrate is dissolved in nitric acid, which converts the uranium in the ore to uranyl nitrate. This is then extracted, dried and heated, whereupon it decomposes to uranium trioxide. Then it is reduced to uranium dioxide by a stream of hydrogen gas. By the action of hydrofluoric acid, uranium dioxide is converted into uranium tetrafluoride—a compound consisting of one atom of uranium to four atoms of fluorine.

Fabrication of Magnox fuel elements at Springfields in Lancashire. (© UKAEA.)

The next step depends upon the type of reactor for which the fuel is intended. For the Magnox reactors, which use natural uranium in metallic form, the uranium tetrafluoride is reacted with magnesium to form uranium metal which can then be cast into rods, machined to size and finally inserted into magnesium alloy cans. Alternatively, for reactors such as the advanced gas-cooled reactor (AGR), which require enriched uranium, the uranium tetrafluoride is converted to uranium hexafluoride (hex) by reacting it with fluorine gas. The hex is then sent to the enrichment plant where the required level of enrichment is achieved as we saw in the last section.

Upon its return from the enrichment plant, hex is reacted with steam and hydrogen to produce uranium dioxide, a powder with ceramic properties suitable for the higher temperature regime inside the AGR as compared with Magnox. The powdered uranium oxide is then compressed into pellets, which are then loaded into metallic cans consisting of stainless steel or Zircaloy, an alloy of the metal zirconium. The cans must be made gas-tight to minimise the diffusion of gaseous fission products into the cooling system of the reactor.

Finally, the fuel elements are assembled into clusters, the size and shape of which depend on the reactor type, and are then transported to the power station.

TYPES OF REACTOR

There is a limited choice of materials used to construct the major components of a nuclear reactor. The limitation arises not only from the suitability of the materials, but also their availability in bulk at an acceptable cost. It is the permutation of the available options that gives rise to the variety of reactor types in use in the nuclear power programmes in various countries. In a particular country, the choice of reactor type is usually dictated by political factors (such as the availability of uranium or access to the appropriate technology) rather than technical or economic factors.

There are three major components of a reactor system for which the choice of material has to be made. The first is the fuel itself. This can be made from natural uranium or from slightly enriched uranium in which the concentration of the fissile 235 isotope has been increased from its natural level of 0·7% to about 2–3%. The second component is the moderator, whose job is to slow down the neutrons from the fission process so that they are better able to sustain the chain reaction. Here the requirement is for a light substance, to give efficient moderation, with a low propensity for absorbing neutrons. In practice there are three choices—graphite, which is a crystalline form of elemental carbon, ordinary water or heavy water, which contains deuterium, a heavy isotope of hydrogen. Some more exotic substances have been suggested but, although their properties may be somewhat superior to those listed, they would be prohibitively expensive to use in practice. The third major component is the coolant, which has to transport the heat away from the reactor core to raise the steam to drive the turbines. Here there are four practical choices: helium or carbon dioxide gas, ordinary water or heavy water.

Let us now consider how these various options are combined in the various reactor types in operation today.

The Magnox reactor. The Hunterston A power station on the banks of the Clyde in Scotland. (© UKAEA.)

The British nuclear programme has been based on gas-cooled, graphite-moderated technology. The first reactors to be built were of the Magnox design. The fuel is in the form of metallic rods of natural uranium encased in cans of a magnesium alloy (from which the reactor type gets its name). The moderator consists of blocks of high-purity graphite and the heat from the fuel is extracted by a stream of carbon dioxide gas which is pumped through the system at high pressure (up to 30 atmospheres). The core is contained in a concrete pressure vessel some 14 feet thick, which also acts as a biological shield to protect the operators from radiation, though some of the earlier models had a 5 inch thick steel pressure vessel mounted inside a thick concrete shield.

Because the Magnox uses natural uranium, the purity of the materials is important in the maintenance of the chain reaction, as also is the geometrical arrangement of the core. Similar considerations also dictate the choice of alloy for the canning material. The magnesium alloy chosen has a low absorption cross section for neutrons. Unfortunately it also has a relatively low melting point, and this limits the core temperature to about 500 °C, giving a maximum gas output temperature of about 400 °C. It is this temperature which determines the thermal efficiency of the station, that is the percentage of the heat energy generated in the core which eventually emerges as electrical energy. This is governed by the second law of thermodynamics, and the higher the temperature the higher the efficiency of conversion. For a Magnox reactor, the thermal efficiency is about 30%. One advantage of the Magnox reactor is its high output of plutonium to fuel a possible fast reactor programme in the future. More immediate advantages, proved over twenty years' experience of operation, are its high reliability and excellent safety record.

There were no exceptional problems in building Magnox reactors, but a few worries have appeared since. The first, which was noticed quite early on in the programme, was the observation that certain mild steel nuts and washers were corroding faster than expected at the highest coolant temperatures. It seems possible that the nature of the steel required had not been specified with sufficient precision, and that the original intention had been to use stainless steel.

In order to slow down the rate of corrosion, so that these nuts and washers will survive for the operating life of the stations, it has been necessary, in some cases, to reduce the maximum output temperature by some 20–30 °C, thus reducing the maximum output of the station by 10%. This derating (reduction of the maximum performance) of some Magnox stations has lead some people to denigrate the Magnox reactor programme. To be fair, one must point out that most power stations—nuclear or not, in Britain and elsewhere—are, for technical reasons, never run at their maximum design output.

A second problem is more recent: on a routine inspection small cracks were found in some of the welds between the restraining bars and the bellows in the coolant gas circuit. Though not a major problem, it did mean that the stations had to be closed down for a while for repairs. Some of these stations are getting on for twenty years old and so it is not surprising that minor repairs should be necessary from time to time. Generally speaking, our Magnox reactors are in good shape, and there is no reason to suppose that they should not be capable of delivering power for many more years to come. They produce some of our cheapest electricity, and with rising fossil fuel costs are likely to prove even cheaper in the future. There is a strong incentive, therefore, to keep them on stream for as long as possible.

The advanced gas-cooled reactor (AGR) is, as its name implies, a souped-up version of the Magnox. It was the next logical progression in the development of graphite-moderated gas-cooled reactor technology. The problems encountered with the British AGR programme, which were mainly institutional and political in nature, will be discussed in a later chapter. Here we are concerned with the technical details of the system.

The major improvement is the use of enriched rather than natural uranium for the fuel. This enables the core size to be reduced with a subsequent reduction in capital cost. Also, the fuel is made of uranium oxide rather than uranium metal, and is encased in stainless steel rather than magnesium alloy. Thus the fuel elements can be run at a much higher temperature (about 750–800 °C), giving a higher coolant gas output temperature of about 650 °C compared with only 400 °C from Magnox. This increases the thermal efficiency of the station to over 40% compared with 30% from Magnox stations. Like the later Magnox stations, the core and coolant are contained in a concrete pressure vessel some 14 feet thick, which also acts as a biological shield.

It had been hoped to be able to refuel the AGR on load, as is Magnox, and hence increase the operational efficiency of the station. So far this has not been achieved, but progress is

The advanced gas-cooled reactor (AGR). Hinkley Point B power station on the Somerset coast. (© CEGB.)

being made and refuelling on partial load is expected to start shortly. The main difficulty has been that of lowering the new fuel elements into their channels in opposition to a stream of high-pressure coolant gas. Although severe technical problems have had to be overcome in the present AGR programme, their solution has built up a fund of valuable construction and technical experience for the British nuclear industry to which operating experience is now being added.

Research is going on to push gas-cooled reactor technology to produce even higher coolant outlet temperatures. Such high-temperature reactors (HTRs) could theoretically reach gas outlet temperatures of 800–1000 °C, leading to a significant increase in thermodynamic efficiency. One major problem is that conventional steam turbine technology could not handle such high temperatures because of the corrosive properties of steam at these temperatures. Thus a major redesign of the steam turbine is required, or some other method devised for converting the heat into electricity. This problem is considered by some people to make this line of development inherently unattractive compared with other reactor designs. An alternative application of the high-temperature gas is to use it directly rather than for electricity production; for example, for steelmaking, which currently uses vast amounts of fossil fuel, if a high enough temperature can be achieved.

The high-temperature reactor (HTR). The Dragon Project at Winfrith in Dorset. (© UKAEA.)

Research on HTRs was carried out for many years in Britain in the aptly named Dragon Project at Winfrith in Dorset. It used helium gas at over 40 atmospheres as its coolant, and temperatures of 750–800 °C were achieved. Several other countries also participated in this project, which was eventually closed down in 1976, but work is continuing in Germany and the USA.

The American nuclear programme has been based on water-cooled and moderated reactor technology. The most popular version of this type has been the pressurised water reactor (PWR) and indeed this model has been adopted as the basic reactor design in most countries with nuclear programmes.

The PWR also uses enriched uranium fuel in oxide form clad in Zircaloy. Water kept under sufficient pressure (about 150 atmospheres) to prevent it boiling is pumped through the reactor core to extract the heat, and also to act as the moderator. The coolant output temperature is about 320 °C, which leads to a thermal efficiency for the station of about 32%. The core and coolant are contained in a steel pressure vessel up to eleven inches thick surrounded by a concrete biological shield. All this, together with the pipes leading from the high-pressure cooling circuit, are placed in a concrete 'containment' building capable of containing within it any radioactivity should there be any leak in the coolant system.

The pressurised water reactor (PWR). The Robinson SE Plant at Hartsville, South Carolina, USA. (© Carolina Power and Light Co.)

An alternative form of light-water reactor is the boiling water reactor (BWR). Water at about 70 atmospheres is pumped through the core where about 10% of it is converted into steam which is fed directly to the turbines. It is thus a 'direct cycle' system and has sometimes been referred to as a 'nuclear kettle'. It has not proved as popular as the PWR.

CANDU. The Pickering power station on the shores of Lake Ontario in Canada. (© Ontario Hydro.)

Let us complete our survey of the thermal reactor systems in use in the world today by describing those which use heavy water. The most successful has been the CANDU system, which was developed in Canada. It uses natural uranium in oxide form, canned in Zircaloy, as its fuel. The core of the reactor consists of a calandria (a tubed tank) containing the heavy-water moderator at low temperature and pressure. Pressure tubes containing the fuel elements pass through the calandria tubes. Heavy-water coolant at high pressure (110 atmospheres) is pumped through the pressure tubes to extract the heat from the fuel. The pressure is sufficiently high to prevent boiling of the coolant, which passes from the core at a temperature of 290 °C to a heat exchanger in which ordinary water is raised to steam to drive the turbines. There is thus no pressure vessel as such, the high-pressure coolant being contained in a number of individual tubes. Failure of any one of these would be serious but not catastrophic. The reactor is housed in a thick-walled concrete containment building designed to withstand up to nearly half an atmosphere of overpressure to contain the evaporated coolant should such an unlikely failure occur.

The Canadians decided quite early on that this was the best system for them. They have stood by their decision and now have a reactor system which is at least as good—some would say better—than any other in the world and on which they have amassed considerable experience both in construction and design.They have also been successful in selling a few reactors to foreign customers.

The steam-generating heavy-water reactor (SGHWR) at Winfrith in Dorset. (© UKAEA.)

The steam-generating heavy-water reactor (SGHWR) was developed in Britain and a 100 MW prototype has been running successfully at Winfrith for many years. As in CANDU, there is a calandria containing the heavy-water moderator and pressure tubes for the fuel. The fuel is again uranium oxide clad in Zircaloy, but it is slightly enriched (1–2% uranium-235). This enables the thickness of the pressure tube walls to be increased to give extra safety. The coolant is ordinary water pumped through the pressure tubes at high pressure where it is raised to steam and passed to the turbines. So it is a 'direct cycle' system similar to the BWR.

All the reactors we have described are classified as 'thermal' reactors, because they use moderators to slow down the neutrons to enable them to sustain the chain reaction more efficiently. In chapter 2 we also discussed the 'fast' reactor which is under development in a number of countries and which will, when deployed, enable us to make much better use of our uranium resources. As we discussed it in some detail in that chapter, there is no need to do so again here.

DECOMMISSIONING REDUNDANT STATIONS

A question that is often asked is what shall we do with nuclear power stations when they come to the end of their working lives. By then they will be intensely radioactive, or at least parts of them will be, so their demolition will be a somewhat more difficult task than that of a redundant coal- or oil-fired power station. Clearly it would be unsatisfactory in the long term just to entomb them and leave a legacy of radioactive mausoleums to our descendants. The question of what to do with old nuclear power stations is therefore a serious one, and one that must be faced.

How long should a nuclear power station be expected to last? The accountancy life of a nuclear power station—the period of time over which the capital investment in the plant is written off, and is used to calculate the cost of electricity produced by the plant—is twenty years. In practice a nuclear power station is expected to operate for very much longer than that as there is no technical reason why it should not do so. Because the running costs are so very much lower than those of coal- or oil-fired stations, there are strong economic reasons for keeping nuclear power stations on stream for as long as possible. The first nuclear power station to be commissioned, Calder Hall, has now been operating for nearly a quarter of a century, and there are no technical reasons why it should not continue to do so for many years more. We can thus expect the working life of a nuclear power station to be about forty years, which is about the same as that of other types of power station.

How then would one go about demolishing it after this period of time? The most radioactive component would be the last charge of fuel in the core, but as the reactor is designed for many changes of fuel throughout its working life, there would be no problem in removing this and sending it off for reprocessing as we shall discuss later. Much of the station will consist of turbines, generators, cooling towers (if it has them), containment buildings and buildings housing ancillary equipment of various sorts. These will be just the same as one would find at any other kind of power station. They will not be radioactive, and their removal would be a fairly standard, if rather large-scale, demolition job. That disposes of about 90% or so of the station.

One is now left with the reactor core surrounded by its thick concrete shield. This will be the most difficult part to dispose of because, at least on the inside, it will be intensely radioactive. It would be possible, though difficult, to demolish it carefully and remove the rubble from the site. This is what one would have to do if the site was required for some other purpose. This has, in fact, been done with a small research reactor in the USA. The site was monitored after it had been cleared and it was shown that no trace of radioactivity remained, and that the site had been returned to its original 'green-field' condition, and could then be used, for example, for agriculture or for housing.

One snag, of course, is that the radioactive debris has to be taken and dumped somewhere else, and at least minimal supervision of it would be necessary. It is more than likely, however, that the site will be required for another power station. A preferable solution in such cases would therefore be to leave the rump of the old power station where it is and build the new one alongside it. The residual radioactivity arises from the continued neutron bombardment of the steel components of the core and the inside of the biological shield during the working life of the reactor, and the isotopes so produced tend to have relatively short half-lives. The most significant is cobalt-60, which has a half-life of about five years. By leaving the old reactor core where it is for, say, a further thirty or forty years, this activity will have decayed away to virtually nothing. It can then be demolished without too much difficulty.

The space taken up by the old core is small compared with that occupied by the new station, so it is not likely to be an inconvenience on that score. The biological shield was designed to contain the intense radiation produced in the core when the reactor was operating at full power, so it should have no difficulty in containing the residual radioactivity during the period after it is closed down. It therefore presents no radiobiological hazard to the construction workers or operators working on the new plant. As it is contained within the perimeter fence of the site, provision of the minimal supervision of it over the timescale we are talking about should present no problems. The procedure would therefore appear to be a sound one.

Although it has not yet been necessary, nor is it likely to be for some years, to decommission a nuclear power reactor, the procedures outlined above have been worked out in some detail. There will be an opportunity to try them out shortly when the small prototype AGR at Sellafield is decommissioned sometime in the 1980s. Because of the added complication of residual radioactivity, decommissioning a nuclear power station will cost somewhat more than that of a coal- or oil-fired station. The generating boards are therefore putting aside an annual sum of about £30 million to cover this extra cost.

TRANSPORT OF SPENT NUCLEAR FUEL

The individual fuel rods in the core of a nuclear reactor have a finite working life of three to four years. After this period of time, fission products have accumulated in the rod to such an extent that they begin to impair its performance seriously. The rods must therefore be removed from the reactor at this stage, and replaced with fresh ones. The spent fuel rods are extremely 'hot'—in the radioactive sense—and special handling gear is needed to remove them from the reactor and transfer them to cooling ponds (they are also quite hot in the thermal sense).

Spent fuel rods being stored in a cooling pond at a nuclear reactor. They reside here for several months before transportation to the reprocessing plant. (© UKAEA.)

The ponds are some thirty to forty feet deep, which is sufficient to shield the personnel from the radiation and also to keep the rods cool. The rods are surrounded by an eerie blue glow, arising from a phenomenon known as Cerenkov radiation which is produced by the passage of beta particles (electrons) through the first few inches of water closest to the rods. Instinctively one expects things that are hot to glow red, and it is perhaps unnerving to the uninitiated to see them glowing blue. The rods remain in the cooling pond for between six months and a year, by which time the short-lived radioactive isotopes—those with half-lives of minutes up to months—will have decayed away.

Test recovery of a spent fuel rod container from a depth of 700 feet. (© UKAEA.)

So far the spent fuel rods have been handled by trained staff and are held on the power station site. The next stage is to transfer them by road or rail from the power station to the reprocessing plant, which in Britain means Sellafield. Even after a year in the cooling pond the rods are still very dangerous—if completely exposed, they could give a lethal dose of radiation to anyone approaching them within a matter of minutes—and so extreme care is needed with this operation. They are transported in massive steel containers weighing up to ninety tons and with walls about twelve inches thick, each containing between two and three tons of spent fuel.

Fire test on a spent fuel rod container. (© UKAEA.)

The containers are designed to withstand the most severe transport accident without releasing any radioactive material into the environment. In order to confirm that the design criteria are sound, full-scale tests were carried out at the Sandia Laboratory in the USA. These included crashing a container on the back of a lorry into a solid, unyielding concrete wall at 80 miles per hour, hitting it with a diesel locomotive at a similar speed and then subjecting it to an intense fire. These severe tests showed that the containers only suffered surface damage and that their integrity was in no way compromised.

Drop test on a spent fuel rod container. (© UKAEA.)

Spent nuclear fuel is but one of many potentially hazardous materials that are transported on our roads and railways in order to sustain our industrial way of life. It is by no means the most hazardous—the risk it presents is no greater than that presented by the transport of, for example, chlorine, petrol or liquefied natural gas—and it is far better protected. Over the years since our nuclear programme first started there have been many thousands of shipments of spent fuel. There have been a few minor accidents, but there has never been a release of radioactivity. Perhaps some day we might take steps to ensure that the transport of *all* potentially hazardous material is handled as safely as this.

Full-scale crash test on a spent fuel rod container. (© Sandia National Laboratory.)

REPROCESSING THE SPENT FUEL

We saw earlier that the fuel rods have a working life of three or four years, after which they are discharged from the reactor and replaced by fresh fuel. The major constituent of the spent fuel is unburnt uranium. This comprises about 96%. About 1% is plutonium, produced by neutron capture by the non-fissile isotope uranium-238, and the remaining 3% is waste material, consisting of fission products (the fragments resulting from the fission of uranium-235) and actinides formed by successive neutron capture by uranium-238.

What then should we do with these spent fuel rods? The worst thing we could possibly do is to throw them away. There are three reasons why this is a bad idea. The first is that the fuel rods are not in a suitable chemical and physical form to contain their radioactive waste component satisfactorily over a long period of time. This is hardly surprising, as they were designed to be fuel rods for optimum use in a reactor rather than as radioactive waste containers. Disposing of them in an untreated form could therefore present an unacceptable and unnecessary risk on future generations.

Spent nuclear fuel being transported to the reprocessing plant at Windscale (now called Sellafield). (© UKAEA.)

The second objection relates to the safeguarding of the plutonium from the unwelcome attention of terrorists. We saw that in the early years after discharge from the reactor the spent fuel rods are intensely radioactive and can only be handled with specialised equipment, and so to a large extent they are 'self-protected' against unauthorised removal. After a few decades, however, the short-lived and hence most intensely radioactive isotopes will have decayed away and the rods will no longer have this self-protection. True, anyone handling them illicitly would get a massive dose of radiation which would almost certainly shorten his (or her) life, but the dose would not be instantly lethal. In time a depository containing spent fuel rods would effectively become a 'plutonium mine', which could attract organisations of evil intent—be they rogue governments or terrorist groups.

The third objection to the disposal of spent fuel rods is that it is a criminal waste of a valuable energy resource. In a world desperately running short of energy that would be clearly undesirable. Only 3% of the spent fuel is actually useless waste. The other 97% is useful fuel that can be recycled.

It is therefore preferable to send spent fuel rods for 'reprocessing'. Reprocessing is a word that has developed emotive overtones in the nuclear debate. It has become shrouded with mystery and superstition. The reality is much less exciting. It is the name given to a rather ordinary though cleverly designed chemical process used to separate out the uranium and the plutonium from the waste products. The former two can then be recycled through the nuclear industry as fuel, and the latter can be packaged into a form most suitable for ultimate disposal.

Windscale (now called Sellafield) on the Cumbrian coast where spent fuel from Britain's nuclear power stations is reprocessed. On the left can be seen the cooling towers of Calder Hall power station, the world's first nuclear power station. The spherical object on the right is the prototype AGR, soon to be decommissioned. (© UKAEA.)

The reprocessing plant which serves the British nuclear programme is located at Sellafield in Cumbria. When the spent fuel rods arrive at Sellafield from the power stations they spend a few more months in cooling ponds before going into the plant. The next step is to remove the canning material; this is done with a mechanical stripper. The stripped cans are radioactive, partly because of activity induced by the high neutron flux in the reactor and partly because of trace amounts of spent fuel adhering to them.

The reprocessing plant at Windscale (Sellafield). (© UKAEA.)

The fuel is then dissolved in nitric acid and the resulting liquid is treated chemically to divide it into three streams, one containing the unburnt uranium, a second the plutonium and the residue containing the waste products. No chemical separation process is 100% efficient, of course, and so trace amounts, perhaps a per cent or two, of uranium and plutonium end up in the waste stream. The chemical technology involved in this process is not particularly exotic, but it differs from other chemical processes in that the working substance is highly radioactive and extreme care has to be taken to protect the process workers by using massive shielding, and to guard against leaks. Because of the high level of radioactivity it is a potentially dangerous process, and indeed minor accidents have occurred from time to time. It is not uniquely dangerous, however, compared with some other chemical processes, and the extent to which the danger can be controlled is witnessed to by the health record of workers on the plant. Not only does this compare well with that of workers on other types of chemical plant, it also compares favourably with that of the public at large.

The separated uranium can then be sent off to be re-enriched and fabricated into fresh fuel rods. The plutonium is at present mostly held in store as fuel for the fast reactors, though some is used to fuel the prototype we already have. The existence of these stocks of plutonium does not commit us to embarking on a fast reactor programme; it could quite easily be used to fuel our existing thermal reactors (possible, but a crying shame). We shall see what happens to the waste in the next section.

Reprocessing technology is most advanced in Britain and France. Both countries have been in the nuclear power business for many years and neither has substantial reserves of indigenous uranium. They clearly have an incentive to use this fuel as effectively as possible; hence their early commitment to reprocessing and their interest in the development of fast reactors. Because of their long experience in reprocessing, both countries are in the business of offering this technology as a service to other countries which have decided to develop nuclear power more recently. Between them, they have contracts to reprocess spent fuel for Japan, West Germany, Italy and Sweden for example. Canada and the USA also have well established nuclear programmes, but as both also have considerable indigenous supplies of uranium the need to reprocess their spent fuel has not seemed so urgent.

MANAGEMENT OF THE WASTE PRODUCTS

There is a saying that 'all that glisters is not gold'. A similar disparity exists between the property of radioactivity and the nature of nuclear waste. The phrase 'nuclear waste' is commonly and quite properly used to describe the waste arising directly from nuclear reaction itself—the fission products and the actinides—as we saw in the previous sections. It is referred to, with some justification, as high-level waste. It is intensely radioactive, particularly in its early years, and would be a potential threat to public health if not handled properly. There is also some other effluent and debris arising from the activities of the nuclear industry, and also from other industries using radioactive materials, which is only slightly radioactive and this is referred to as medium- or low-level waste. It is important to appreciate this distinction between high-level waste and medium- and low-level waste.

Disposal of low-level waste by shallow burial. This is the remains of an accelerator called NIMROD on which the author worked for many happy years early on in his research career. (© UKAEA.)

Low- and medium-level waste consists of things like the metal cans stripped of the spent fuel, cooling water in the storage ponds, or worn-out equipment from the active part of the plant. If the radioactive isotopes are short-lived, then the material is sometimes stored on site for a number of years to reduce the activity still further prior to disposal. The method of disposal depends upon the nature and quantity of material involved. Liquid and gaseous effluents are diluted and discharged to the sea or atmosphere respectively. Solid waste is either buried on land—a shallow trench is quite adequate—or sealed in concrete drums and dumped out at sea. The drums will eventually corrode and crack open, of course, over a period of many years, but the total quantity of radioactive material involved will add a trivial contribution to the naturally occurring radioactivity in sea water. Also, the rate of dissolution in sea water will be extremely slow, thus providing an extra safety barrier.

From Britain we undertake one sea dump each year at a specified spot in the North Atlantic. About 40% of the material dumped is hospital waste arising from medical use of radioisotopes, so we would still need our sea-dumping programme even if we had no nuclear power. The exercise is carried out under strict internationally agreed regulations and is subject to international inspection. Similar strict regulations apply to all discharges of radioactive material to the environment, and because they occur from a limited number of sites, all of which are in the public sector, they can be (and are) enforced. Industry in general has to dispose of large amounts of toxic (non-radioactive) waste resulting from its activities. Government regulations apply here also, but because much of this waste arises from a large and heterogeneous private sector enforcement is more difficult and there have been cases in recent years, albeit small in number, where they have been evaded.

There has been much confusion leading to concern—in my view unnecessary concern—arising from a failure to distinguish between low- and high-level radioactive waste. There was a classic case in a small village in Oxfordshire not far from where I live. The village has a domestic refuse tip, and it was proposed to dump on the tip some slightly radioactive waste from the local hospitals and university departments, mainly swabs and slightly contaminated articles of clothing sealed in polythene bags. The principal isotopes involved had half-lives of a year or so, and the total activity was only equivalent to that of a bagful of luminous watches. Nevertheless the proposal caused an uproar in the village. It was 'radioactive' and it was 'waste'; therefore it was 'nuclear waste' that they had heard so much controversy about, and they did not want nuclear waste dumped in *their* village. A series of public meetings was held in the village, but no amount of rational argument could make the slightest impact upon them and was met with deadpan expressions and suspicious disbelief. They persisted in persecuting the perpetrators of the proposal with all the vigour with which their ancestors had no doubt in previous centuries persecuted witches, and with about as much logic. The District Council could see no valid objection, neither could the County Council nor the Department of the Environment, so the villagers threatened to take their case to the European Court of Justice! In the end commonsense prevailed over witchcraft, but only after a lot of unnecessary hassle.

The discharge of low-level radioactive waste is strictly regulated and has a negligible effect on the environment. With high-level waste it is, or could be, a very different story. So let us now see how that is dealt with.

The material emerging in the waste stream from the reprocessing plant is first reduced in volume by boiling off some of the water in which it is dissolved or suspended. The liquid is then stored in specially designed tanks. The tanks are made of stainless steel and are double walled. They are placed in thick concrete vaults which are themselves lined with stainless steel and capable of containing all the liquid should the tank burst open. There are spare tanks on standby into which the liquid can be pumped should a fault develop in one of the operational tanks. It is important to emphasise the smallness of the quantities involved. Only twelve such tanks are needed to store all the high-level waste from our nuclear programme—both civil and military—from its inception in the late 1940s up to the present day. The total volume of liquid they contain is equal to about that of a four-bedroomed family house. High-level nuclear waste is very dangerous but there is not all that much of it.

Storage tank for high-level nuclear waste in liquid form. The picture shows cooling coils, which remove the radioactive decay heat, being lowered into a new tank. (© BNFL.)

There is a sound technical reason why high-level waste is stored as a liquid during its early years. The isotopes in the waste with very short half-lives—say up to a month or so—will have decayed away during the period in which the spent fuel rods were being stored, but there will still be appreciable quantities of the isotopes with slightly longer half-lives—say from a few months to a few years—which are decaying fast enough to generate considerable quantities of heat. Because the waste has now been concentrated into a small volume, the heat density can be quite large—up to about 15 watts (W) per litre. It is much easier to remove this heat with the waste in liquid rather than solid form. The tanks therefore contain cooling coils through which water is pumped to perform that function. There are several independent circuits—which are redundant in the sense that any one of them is capable of keeping the liquid cool—to give protection against any occasional malfunction.

In the unlikely event of all of them failing at the same time, it would take about a day for the liquid to reach boiling point and several days longer for the tank to boil dry. A further day would be needed for the temperature of the residual solid to rise sufficiently for the more volatile isotopes to be evolved. This allows ample time for remedial action to be taken, for example transfer to an empty standby tank, or external cooling using hose pipes. If action were not taken, the result, although not catastrophic, would certainly produce a serious threat to public health for the population living near the plant, but no more so than in the case of many other types of industrial plant left unattended for many days after they have gone wrong. Many years of experience have shown that the liquid waste tanks are more than adequate to perform the task for which they were designed.

It is desirable to store the high-level waste in liquid form for several years, after which its heat output has dropped by several orders of magnitude. It is clearly undesirable, however, to continue to store the waste as a liquid for the indefinite future. Eventually the tanks will develop maintenance problems, and one has a general feeling of unease about storing such a potentially hazardous substance as a liquid over such a long period of time. A technique was therefore developed at Harwell in the early 1960s to convert this liquid into a solid—a process known as vitrification.

The waste is mixed with standard glass-making material and cast as ingots in a stainless-steel container, coated with a corrosion-resistant alloy. In principle the technique is very simple—man has been making glass for several thousand years after all—but is made more complicated in practice because of the radioactive nature of the substance to be worked with. The Harwell project, known as 'Fingal', was ahead of its time. By the early 1960s the amount of high-level waste that had accumulated was so small that one could not justify at that stage the commercial exploitation of the technique. The feeling at the time was that it would eventually be needed, but not yet, so Fingal was mothballed. Subsequently Fingal has been reborn—under the code name 'Harvest'. A commercial vitrification plant is now under construction at Sellafield, based on the design of a successful plant already operating at Marcoule in France.

Non-active specimens of the solid produced by the vitrification process. (© UKAEA.)

The brief hiatus has been put to good use, however. It has enabled exhaustive tests to be performed on the glass blocks produced from the Fingal project to ensure that the glass has adequate properties to perform the task required of it. Its chemical and physical stability under various conditions were examined, as were the leach rates—the rate at which water would dissolve out the radioactive constituents from their glassy matrix—at a variety of temperatures. It was found that even if the glass were placed in the most hostile natural environment it would take several centuries before the elements (the old-fashioned climatic kind) could release significant quantities of radioactivity from it. By that time the major fission products—those with half-lives of around 30 years or so—would have all decayed away and only the long-lived actinides would be left.

It has been suggested by some competent authorities that there are better solids than glass into which the waste could be incorporated. For example, one could fix chemically the radioactive isotopes in the waste into the crystal structure of an artificially produced rock-like substance. This would be similar in its chemical and physical properties to some of the rocks that have survived in the natural environment under a wide variety of conditions for perhaps thousands of millions of years. Certainly its leach-resistant properties would be better than those of glass, and it would be capable of containing its radioactive component more-or-less indefinitely. The problem is whether or not it can be done in practice. It is one thing to show in the laboratory that one can bind one individual radioactive isotope chemically in the crystal structure of a rock-like substance, but quite another to show that one can develop a practical process to bind all of the 30 or 40 isotopes which constitute nuclear waste. Enormous problems in chemical technology would have to be overcome, but nevertheless the idea is worth pursuing and research along these lines is proceeding in a number of laboratories around the world.

However, the extensive tests that have been performed on the vitrified waste have shown that his material is adequate for the purpose for which it is intended. If it is eventually shown that synthetic rock is better and can be produced practically, then a changeover to this technology may be justified. There is no justification, therefore, for not proceeding to vitrify the stored liquid waste that we have and which is ready for solidification on the grounds that something even better might come along later. As was stated earlier, the British vitrification plant is being built at Sellafield and should be in operation in a few years' time. The French are somewhat further advanced. They have already built their first vitrification plant at Marcoule, and this has been in operation since 1979.

After the high-level waste has been converted into solid it

will continue to be stored at the reprocessing plant for a further period measured in decades. During this period the activity of the waste will decrease still further, as the isotopes with half-lives of up to a few years decay away. This storage will be either in water-filled cooling ponds or in air-cooled concrete vaults. Again one should emphasise the physical smallness of the quantities involved. Even with the largest conceivable nuclear power programme, the land that would be needed for storage is only a few acres. It can easily be accommodated within the present site of the Sellafield plant.

If we now add up the elapsed time needed to store high-level waste in its various forms we see that this takes us well into the next century before we need proceed to the next step. It will be many decades in the future, therefore, before decisions need to be taken as to what that next step might be, but even so considerable thought has already been devoted to the various options. Because of the relaxed timescale we have the luxury of being able to study and compare these options in some detail, a luxury not usually afforded to the waste disposal strategy of many other industries, not to decide whether or not they are feasible (they are all feasible in principle), but to decide which will be the best.

What are the various options? One school of thought is that we should do nothing. That is, we just leave the waste where it is in its concrete vault. It is not going anywhere and at least we will know where it is, and our descendants can, if they feel so disposed, go and look at it occasionally—who knows, they might even find a use for the stuff! It has the minor drawback that it requires a minimum amount of supervision—though that may mean little more than keeping a record of where it is—for a future period extending into centuries. After that, although it will still be radioactive the hazard it represents is no more than that of an uranium ore body.

Others would prefer to see the waste disposed of ultimately in a way that would not even require this minimal amount of supervision. In other words, it would be non-retrievable and could literally be forgotten about. One attractive option is to fire it off into space. It would certainly be practical to do so, and would not be economically prohibitive bearing in mind the relative smallness of the quantities involved. One drawback could be the reliability of rocket technology. The record shows that the failure rate of rockets on launch is about 2%. Would a 2% probability of a malfunctioning rocket plunging down to some unpredictable point on the Earth's surface with a payload of high-level nuclear waste be considered an acceptable risk? Probably not, but it could well be that rocket technology will be so improved by the next century when the decisions have to be taken that such a risk will become negligibly small.

More down-to-earth proposals are to bury the waste a thousand feet down in suitable geological strata, or to place it on or below the sea bed in a deep ocean. Sea disposal depends on the fact that it will take several centuries for the encased glass blocks to corrode sufficiently for their radioactive content to start to seep out, by which time the fission products will have decayed away, and some millenia after that for them to dissolve. This rate of dissolution is sufficiently slow that complete mixing with the sea water will be achieved, and the amount of radioactivity contributed to the sea water from the long-lived actinides will be negligible compared with the amount of radioactivity in sea water from natural sources. At the present time international law prohibits all but the discharge of low- and medium-level waste into the sea, and would therefore have to be changed if it were decided that this was the best option for the ultimate disposal of high-level waste. There is no technical reason why this change should not be made, but one can easily see it giving rise to interminable political wrangling.

For disposal on land, clay, salt or hard rocks such as granite would seem to be the best geological strata to act as the host material, and research programmes have been carried out in this and other countries to investigate their properties in more detail. Significant properties include the heat conduction of the material *in situ*—even after several decades in storage the waste will still be emitting some heat—and permeability to water. Geological research has been carried out at a site at

Caithness in Northern Scotland, and studies of heat transmission in granite have been performed in Cornwall, where there is a long tradition of hard-rock mining. Also, the clay deposits under Harwell have been examined as a possible host for medium-level waste.

Geological research being performed to check the suitability of granite as a host material for the ultimate disposal of high-level nuclear waste. This is a drilling rig at Altnabreac in Caithness, Northern Scotland. (© UKAEA.)

With this option, high-level waste would be buried about a thousand feet or so below the surface. One site will be required at some time in the early part of the next century, with possibly a second one towards the end of the next century. This is another consequence of the physical smallness of the quantities involved. The individual blocks of waste will be surrounded by a few feet of a suitable packing material with appropriate 'ion exchange' properties that would enable it to absorb the bulk of any radioactive material that might be leached out of the glass.

Thus there are four physical barriers between the waste and man's environment—the glass itself, its stainless-steel container, with its corrosion-resistant alloy coating, the packing materials and a thousand feet of rock. On top of this, one chooses a site which throughout geological history has not had any ground water anyway, this being the only credible medium through which the radioactivity could find its way to the surface.

It is difficult to find any fault in principle with any of the waste disposal options that have been suggested. Some critics have nevertheless demanded that at least one of them must be 'demonstrated', though they are never quite clear as to what they mean by a demonstration. Earlier on we discussed the natural reactor that was discovered at Oklo in West Africa. We saw that the waste products, although completely unpackaged, had barely moved in 1800 million years: that is a pretty good demonstration!

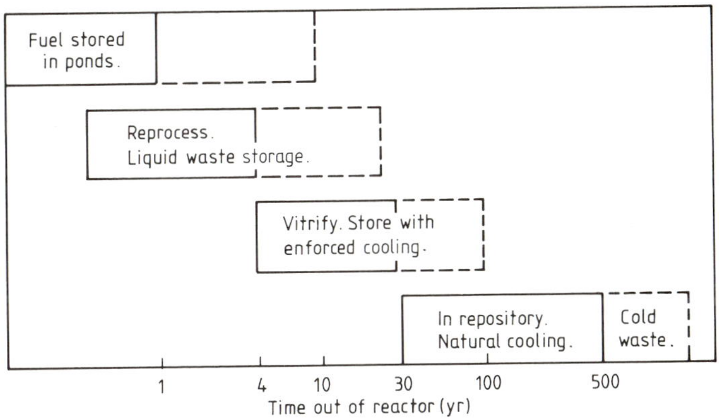

Timescale of storage and disposal of highly active waste. (Courtesy of UKAEA.)

The radiobiological consequences of the responsible management and disposal of high-level nuclear waste are trivial compared with those of the natural radiation background. We shall be discussing them further in a later chapter.

CHAPTER 4

Nuclear Power in the Energy Scene

Today, most of us in the developed countries enjoy a standard of living beyond the wildest dreams of our forefathers. We live in homes with a level of comfort that in earlier generations was available only to a privileged few. We have access to a wide range of manufactured goods, at prices which most people can afford. The basis of our high standard of living rests on our industrial development over the last century or so, and that industrial development has been possible because of the availability of a cheap and abundant source of energy.

Britain, through taking advantage of her rich resources of coal, was the first country to industrialise. The process started towards the end of the eighteenth century, and rapidly gained momentum in the nineteenth. Other countries, in particular the USA and Germany, were not long in following suit. Because it is comparatively recent, it is perhaps easy to overlook the significance of the industrial phase of man's development. For the first time in history the ordinary man had the opportunity to escape from grinding agrarian poverty. The pages of history tend to be dominated by the lives of the wealthy minority of previous generations. Look a little deeper and we find that the common life of ordinary folk in pre-industrial times was hard, brutish and short.

The benefits of industrialisation were not immediately apparent to the ordinary man. It took some time for society to adapt, and to develop the mechanisms which would ensure a more equitable distribution of the newly created wealth. This is a process which continues to the present day and is by no means complete. Nevertheless, created wealth there was, and by comparison with earlier generations stretching back to the birth of our race, wealth in incredible abundance. And it all depended on energy, driving concentrated and controllable power sources that man had developed which multiplied by many hundred-fold the power he could develop from his puny frame.

ENERGY IN MAN'S DEVELOPMENT

The only power source available to primitive man was that of his own muscles. With this he had to gather or hunt for his food (and, after the invention of agriculture, till the land), build his shelter and move himself around. All his efforts had to be devoted merely to his survival. After he had passed the peak of his physical strength in his mid-twenties, his chances of doing that decreased with each passing year.

Eventually he came to realise that although he might be the most intelligent creature on Earth, he was by no means the strongest. He gradually developed the skill of using his superior intelligence to train larger and stronger animals to work for him. This enabled him to till more land, and hence grow more food, to move larger pieces of wood and stone, and hence build a more substantial shelter, and by riding on their backs to be able to travel much greater distances.

Thus far, man's energy requirements were in tune with the annual cycle of nature. The basic 'fuel' he was using was the food eaten by himself and his animals. This came from each year's crop. Indirectly he was using the power of the Sun, and averaged over a year he was in balance with nature: the energy he could collect in his annual crop supplied his energy needs for that year.

Then man discovered fire. He had been familiar with fire for a long time, of course. He had seen the devastating effects of forest fires and the threat they posed to the existence of his community. Over the generations man had developed a healthy respect for, indeed a fear of, fire. It was a major step in his advancement when he learned to control this fearsome phenomenon and turn it to his advantage. With it he could protect himself from predators, keep himself warm through the long winter months, cook his food, thereby making it more digestible, fire clay to make building bricks and domestic utensils and eventually smelt the ores of the metals and hence avail himself of a new technology providing him with tools, implements and weapons which were infinitely superior to any that were available previously.

The fuel to feed his fires came almost exclusively from wood, obtained from trees in the ample forests that surrounded him. The use of wood as a fuel marked a significant change in man's relationship with his environment. Whereas man had previously been in balance with the annual cycle of nature, he was now for the first time in a position to use resources at a rate faster than that at which nature could replenish them. It takes several decades, perhaps a century, to grow a tree. Man could chop it down and burn it in a matter of days. Whilst his numbers were small, and his demand for fuel was modest, his effect on the natural world was small. Slowly, however, over the centuries the sophistication of his civilisation increased. This caused an increase in his *per capita* demand for fuel and also, as this enabled him to support a larger population, to an increase in his numbers as well, thus multiplying further his total energy demand. So his demand for wood as a fuel grew inexorably over this period. He systematically deforested vast areas of the planet.

It was, in fact, an eventual shortage of wood in the sixteenth century which led man to start using another fuel—coal. Because of the noxious fumes emitted when it was burned, this innovation led to opposition from the environmentalists of the day, although they did not at that time label themselves by that modern title. Indeed, in late Tudor times there was a prohibition on the use of coal in London for that reason.

Coal is formed by the action of heat and pressure on long dead living matter that has become buried under the Earth. For that reason it is sometimes referred to, quite correctly, as a fossil fuel. It is the fossilised remains of what lived and grew on the Earth aeons ago. The natural processes involved in producing coal must operate over a long timescale—a timescale measured in millions of years. This is so long compared, say, with a human lifetime that for all practical purposes it is a nonrenewable source of energy. When man started mining coal he was quite literally beginning to dig into his energy capital.

Coal is a very concentrated energy source and, when man had developed the techniques for mining it in bulk, was available in large quantities. A capital resource it might be, but man made good use of this accumulated capital. He used it to fuel the industrial revolution which eventually, as we saw, led to a dramatic increase in the living standards and material wellbeing of his rapidly increasing population.

Towards the end of the nineteenth century man discovered another fossil fuel—oil. For some time prior to that farmers in Pennsylvania had noticed an obnoxious black sticky substance oozing up into their fields, ruining what would otherwise be good arable land. It was a nuisance which threatened the economic operation of their enterprise. Once its potential came to be realised, however, this nuisance became transformed into a valuable asset. Like coal, oil is a concentrated form of energy, but being a liquid rather than a solid it is much more flexible and much easier to win from the Earth, transport and use. So shortly after its initial discovery, demand multiplied rapidly, and the scramble for further and larger supplies began. Vast reserves were discovered in the USA, particularly in Texas, a name that for many years was synonymous with oil. After the Second World War, the enormous oilfields of the Middle East were discovered and developed. Such was the growth in the demand for oil that by 1960 it had replaced coal as the principal fuel of the industrial world.

Not only was oil a versatile and convenient fuel, it was also incredibly cheap to produce and deliver to the market place. It is obvious why this should be so. Unlike coal, there is no need to sink expensive mines and send down armies of men to dig it out laboriously. Nature has obligingly collected it in vast pools and all man has to do is pump it out. Indeed, in some fields it actually comes out of its own accord as it is under pressure. Production costs were only a few cents a barrel. In fact they still are only a few cents a barrel. The other twenty dollars or so on the market price arises from the royalties and taxes levied by the governments of the producing and consuming countries, respectively, for reasons that we shall discuss later.

This cheap and abundant supply of oil fuelled man's industrial society as never before. During the 1950s and 1960s economic growth reached unprecedented levels. It produced material standards of living which, particularly to those who remembered the deprivation of the years of the Second World War, seemed like a dream come true. It went on for so long that we really began to believe that this was the new norm—that economic growth would continue at this rate—that we had solved the problem of poverty and that things must inevitably get better and better. We were to be sadly disappointed. In 1973 we realised that the era of cheap energy was over. During these two decades the price of oil remained more-or-less constant in money terms, which means that in real terms (i.e. allowing for general inflation) it actually fell. In the autumn of 1973 it caught up with a bang.

During these years man also made extensive use of a third fossil fuel—natural gas. It had collected over aeons in pockets under pressure in suitable geological strata. As with oil, it is very cheap to extract. Being a gas, although it can be liquefied, it is less easy to transport round the world than oil. Although there is some international trade in natural gas, it is mainly used in its country of origin, providing several local bonanzas on top of the general world bonanza provided by cheap oil.

More recently, man has started to exploit another of his supplies of energy capital—the metal uranium. This metal, along with the other heavy elements, was created by the forces that formed the Earth and the Solar System some four and a half billion years ago. So there can only be a fixed amount of it on the planet. There are no natural processes on Earth to increase the amount, or to replenish that used by man. In fact, because it is radioactive, the total amount is actually decreasing slowly, but as the half-life is long even compared with the period of man's development that is irrelevant to our story. Like the fossil fuels, it is a capital reserve—once used it is gone for ever. However, uranium is the most concentrated form of energy that man has yet discovered—something like a million times or so more concentrated than even the fossil fuels—so a little of it goes a long way.

We have seen how man has developed over many millenia, and more rapidly over recent centuries, and yet more rapidly still over the last few decades, from primitive beginnings as little more than an animal amongst other animals to his present level of material comfort. To do that he has needed increasing amounts of energy in ever more concentrated form—and to get that he had to dig deeper and deeper into his reserves of energy capital. Anyone who lives on capital knows—and the rest of us can imagine—that one cannot go on this way indefinitely. Sooner or later one will end up in a crisis, and one doesn't have actually to run out before one begins to feel the pinch.

THE IMPENDING CRISIS

What energy crisis? There is enough coal left in the world to supply our needs for more than three centuries. Uranium reserves are roughly comparable. Using thorium as a means for breeding nuclear fuel we could meet our present level of demand for energy for a millenium. Then there is nuclear fusion, using deuterium from the vast oceans of the world. There is a further millenium's worth, and more. Man's energy capital might well be finite. No doubt it will run out sometime. But there seems to be enough for a few millenia yet. In addition there are forms of energy income arising principally from the daily flow of energy from the Sun. We might not be able to harness their full potential, but that potential is vast and it is constantly being replenished. So where is the problem? Again I ask, what energy crisis?

Actually we don't have an energy crisis. What we do have is an oil crisis. We saw earlier how, because of its cheapness and versatility, oil very rapidly became the major energy source of the industrialised world. One could say we became hooked on oil. However oil, as we saw, is a nonrenewable source. Surely we couldn't be stupid enough to let its reserves drop to such a level that constraints on its supply could seriously affect our economic wellbeing without taking adequate steps to ensure that other energy sources would be ready in time, could we? Yes we did!

The first ominous signs became clear by about 1950. At about that time the rate of increase in demand began for the first time to outstrip the rate at which new discoveries were being made. We failed to heed this warning: on the contrary, awash with cheap and abundant oil from the newly developed fields of the Middle East, we forged ahead with its profligate use as if we thought that the energy bonanza would go on for ever. Future historians will look back on the decades of the 1950s and 1960s with incredulity. To make prodigious use of oil while it was cheap and plentiful was perhaps excusable. But to have allowed ourselves to have become so besotted with a single energy source—obviously nonrenewable, and of limited long-term availability—that we failed to prepare ourselves for the inevitable future shortages is totally inexcusable.

During these two decades we came to regard coal as 'yesterday's' fuel, superseded in the inexorable flow of economic progress by cheap, plentiful and convenient oil. We systematically began to run down our coal industry. Some of our European neighbours, particularly France and Belgium, did so far more effectively than we did in Britain. They succeeded in running down their coal industries to past the point of no return. In fact when the last Belgian mine closed the authorities made quite an event of the last miner coming back to the surface. Fortunately we were not quite as efficient as that. We have a fighting chance to retrieve the situation.

The miners, of course, were bitterly opposed to the rundown of their industry. They claimed that it was essential to keep the coal industry alive, because sooner or later we would again need substantial supplies of coal. Nobody believed them, but in the event they turned out to be right. Had we taken the trouble to step back from the picture and look at the energy situation in its true perspective it would have been obvious that eventually they were bound to be right.

Fortunately the correctness of their claim is now generally recognised. The National Coal Board has an ambitious investment programme—the Plan for Coal—designed to raise output from its present level of 120 million tons per year to 180 million tons per year by the end of the century. By then we will require coal to take over some of the roles now performed by oil, not only as a fuel for providing process heat in industry, but also as a raw material for the production of liquid fuels for transport, a feedstock for the petrochemical industry which produces the plastics and other synthetic materials which feature so largely in our present way of life, and for the production of substitute natural gas when our supplies from the North Sea start to decline. And of course we shall still need it to produce a large fraction of our electricity.

After decades of neglect, the Plan for Coal is indeed an ambitious programme. There is no guarantee that it will succeed—there is already some apprehension that it might not. Let us hope that it does, because by then we are likely to need every last shovelful.

The coal industry was not the only casualty of the era of cheap oil. In the years following the Second World War, Britain embarked on a vigorous programme to develop civil nuclear power. However, when cheap and abundant supplies of oil began to flow from the rapidly developing oilfields of the Middle East the programme gradually lost momentum, and eventually slowed down almost to a standstill. It became increasingly difficult to persuade the CEGB and the other electricity generating authorities to invest in nuclear power stations when they could at that time produce electricity more cheaply from oil. Observers of the nuclear debate should note that the CEGB has not always been a strong supporter of nuclear power. Throughout the two decades of the 1950s and 1960s its enthusiasm for this new technology could hardly be described as anything more than cool. The CEGB's 'conversion' was recent and dramatic, dating as we shall see from that fateful autumn of 1973. Once again, as a nation we failed to look ahead and plan for the future.

In the late 1940s Britain established a modest but respectable research programme aimed at developing wind power as an energy source. Again, when the cheap oil started to flow the programme ground to a halt and was eventually abandoned in the early 1960s. It was not until the oil crisis of 1973 that we realised our mistake and put steps in hand to attempt to retrieve the situation.

I well remember visiting one of the leading lights of that earlier windpower programme, to discuss the renewed interest in the subject. With that look in his eyes that could only be described as 'I told you so', he opened his filing cabinet to retrieve the threads of his work that he had been obliged to drop some twenty years previously. Regrettably he was to die of leukaemia a few months later, but at least he had the satisfaction of knowing that his earlier work to which he had devoted several years of his life, apparently worthlessly, had not been in vain.

The halcyon days of the 1950s and 1960s were not without their occasional interruptions in the flow of oil from the Middle East—interruptions arising from the troubled political situation in that part of the world that occasionally boiled over into war. Britain has had a long history of association with the Middle East—not all of it happy—extending back to her imperial days. The impression had become firmly established, particularly in the minds of the Arab residents of the area—on whose land the bulk of the oil reserves are located—that when trouble arose, somehow or other Britain was behind it. It wasn't always true, of course, but that was the impression. So the habit was formed, almost as a political reflex action, that when hostilities broke out in the Middle East, the Arabs cut off oil supplies to Britain.

This happened in 1956 during the Suez Affair, in which we were quite definitely involved, and there was petrol rationing in Britain for a limited period. It happened again in 1967 during the Six Day War, in which we were certainly not involved. This time, even though there was simultaneously a civil war going on in Nigeria—our next largest supplier—and oil was unavailable from there also, our shortfall was made up from the oilfields of the USA and Venezuela, such was the excess of production capacity over demand at that time.

Yet only six short years later, when it happened again with the Yom Kippur war of 1973, the whole situation had changed. Demand in the USA had risen so dramatically that she had moved from being a net exporter to a net importer of oil. Oil had changed from a buyer's to a seller's market. No longer could suppliers outside the Middle East come near to matching the then level of demand unaided. The Arab States, who had long felt aggrieved at the low royalties they had been getting for their oil, seized their opportunity. They formed a cartel called OPEC (Organisation of Petroleum Exporting Countries) and collectively raised the price to four times its previous level—an action that triggered off the worst economic depression that the Western World had seen since the 1930s. The oil still cost the same to produce, of course, so this vastly increased revenue caused an inflow of cash into the oil-rich countries of gigantic and unprecedented proportions. It became clear that oil kept in the ground was of more value to them than money in the bank in quantities greater than they could reasonably spend, so most of them reduced the output from their fields, causing further problems for the industrialised world.

The events of 1973 and subsequent years are indicative of the fact not that oil is in imminent danger of running out but that the relationship between supply and demand is now such that it is the supplier who calls the tune. They are a clear sign that oil is entering its final phase as a plentiful resource; they are a timely reminder that oil reserves will not last forever. Certainly we will continue to find new fields in different parts of the world, but at our present rate of demand we would have to make discoveries of fields the size of Texas *every year* just to keep up. Certainly we are not doing that, nor are we likely to. Clearly the writing is on the wall.

How long do we have to put our energy house in order, before a difficult situation become desperate? Predicting the future is a hazardous business. One can, however, identify certain trends which might give us some idea of what the future has in store. One trend, as we have seen, is the perilous state of world oil supplies, a condition that can only get worse as the years roll by. It will not be many years from now when world oil output will start to decline, even without political constraints.

Another clear trend is in the area of world development. Very few nations enjoy anything like the same standard of living that we do. The discrepancy between the world's rich and the world's poor is enormous. It is a disgrace that it has continued for so long. Furthermore it is an unstable situation, which cannot and must not be sustained for much longer. We used to refer to the poorer nations of the world as 'backward'. As that could be interpreted to imply that they were in some way basically inferior, that was rather insulting so we have stopped saying it. So we changed it to 'underdeveloped'. That also was rather insulting, because it implied that they had been less smart than we had been in developing their natural resources. A more accurate description is 'developing', because that is just what many nations in the Third World are now doing. No longer are they prepared to survive on handouts from the rich nations, to live off the crumbs from the rich man's table. They are building for themselves their own industrial base from which to create their own wealth.

They won't necessarily follow the same path of industrial development that we did—times have changed and circumstances are different—but developing they certainly are. There is another thing that is certain: that development is going to require that they demand a much larger share, they could quite reasonably claim a fairer share, of the world's resources. One of the resources, undoubtedly, will be energy. To gain some glimpse of the magnitude of the phenomenon that is about to befall us, let us look at how energy is now used in different parts of the world. To do that we can define a convenient unit of energy, tons of coal equivalent (TCE), that is the amount of energy used converted to its equivalent in tons of coal. It will not all be used as coal, but it is a convenient way of referring all fuels to a common base.

It will not surprise the reader to learn that the country which uses the largest amount of energy per head is the USA—about 12 TCE per person per year. For the countries of Western Europe, including Britain, the rate of consumption is about 6 TCE per person per year. In a typical Third World country, it is only about 1 TCE per year. As the Third World countries develop, and start to demand more energy to power their expanding industrial base, the pressure on world energy reserves is bound to be enormous, almost irrespective of any expansion or otherwise in the countries already developed. In particular this demand will be for oil, since this fuel requires the minimum investment for its use. And to compound the problem, we must remember that there are a lot more of them than of us.

So we can identify two trends: an inexorable increase in world energy demand, driven by the long-awaited development of the Third World, and declining resources of oil. Inevitably these two trends must come into conflict, probably some time around the year 2000, unless definite steps are taken to prevent it. We have barely twenty years to sort ourselves out. Bearing in mind how long it takes to research, develop and deploy new technologies, it might be touch and go.

But what about us in Britain? Are we not in an uniquely favourable position, particularly with our North Sea oil? The discovery of oil, and gas, in the UK sector of the North Sea was undoubtedly the best stroke of luck that we have had in many a long year. But if we allow it to beguile us into a false sense of security, it may also be the cause of our downfall. It is expected that the production of oil from the North Sea will start to decline by the early 1990s, and that by the year 2000 Britain will again become an energy-importing nation—we will be thrown back into the world market at a time when we can expect the world oil situation to be desperate.

So North Sea oil provides us with a breathing space, and no more than that, in which we can restructure our energy resource base for long-term viability in a fairly orderly manner. It is an opportunity that we cannot afford to miss. We must beware of the fate of our neighbours in Holland. They too had a lucky windfall, with the discovery of large quantities of natural gas in their sector of the North Sea. They used it to fuel an unprecedented industrial expansion, and with the wealth thereby created they established one of the most generous welfare states in the world. Now their gas supplies are in decline, how will they continue to support their excellent social programmes? There is a marked reluctance amongst people in Holland to turn to nuclear technology but they may find that they have no option. The Dutch experience is one that we in Britain should note with some apprehension. We have about twenty years. Let us not waste them.

THE NEED FOR ENERGY CONSERVATION

We have just come through an era of cheap energy, based mainly on abundant low-priced supplies from the Middle East. It was an era that lasted for just over two decades. We have developed some bad habits because of it. We have become profligate in our use of energy. After all, why should we bother to use energy wisely when it is so cheap and plentiful? But that era has come to an end, and with its demise we must now adopt a different attitude towards energy and the way we use it. As well as changing to alternative supplies of energy, which must be more expensive than oil was or we would have been using them more extensively anyway, it is just as important, indeed it could be more important, to adopt ways of using our precious supplies of fuel more efficiently.

It does not necessarily mean that we have to reduce our living standards, but it does mean that we are going to have to learn to get the maximum use from each ton of fuel that we use. There is a lot that can be done. Energy conservation must become the watchword of the new age in which we find ourselves. It is one of the major weapons we must use in our fight to avoid a catastrophic energy situation developing.

Much can be done without our having to spend any money. Good housekeeping, in the home, the office, the factory and in the way we use our cars, can lead to worthwhile savings of energy and it doesn't cost us anything—simple things like switching off lights and equipment when not in use, not having the heating turned up too high, drawing the curtains after dark and preventing draughts. It is just a question of establishing the habit. Do we really need to drive as fast as we do? There are occasions when we are in a hurry and we bomb down the motorway at seventy, but there are many more occasions when those extra few minutes saved are not essential. Petrol consumption rises dramatically when the speed increases from fifty to seventy miles an hour: the difference is typically forty compared with thirty miles to the gallon. Think about that next time you are pumping pound notes into your petrol tank.

To go further than this requires some investment of capital. It means buying some hardware, or modifying existing hardware, on the basis that the costs involved in doing that will be more than recovered from the reduction in fuel bills that result. That raises two questions. First, does the technology exist, i.e. is the hardware available on the market? Second, is it economic, i.e. will the money saved on reduced fuel bills cover the cost of buying and installing the hardware in a reasonable period of time? The answer to both questions is yes in most cases. I have even heard it claimed that it is always cheaper to save a unit of energy than to produce a new one. Although it is often true it is not universally true. For example, I was working on a scheme some time ago to recover the heat from used bath water by using a small heat pump to pump the heat back into the hot water tank. It sounded a good idea, but there was no way that the few pence per year one would save on heating bills could effectively pay for the purchase and installation of the equipment. Most energy conservation technologies make sound economic sense, but there are some that do not, and so the energy consumer must find out the facts case-by-case and exercise his judgment.

Almost invariably loft insulation in the home is economic, and so also is cavity-wall insulation. Double glazing, on the other hand, is rather marginal, and is only worth doing to give another benefit such as sound insulation. In industry also there is much to be done to insulate buildings, many of which

Loft insulation being installed. This young man should be wearing a face mask, as the fibre can be dangerous if inhaled. (© UKAEA.)

were put up in the days when energy was cheap. There are also many cost-effective ways of improving the energy efficiency of the industrial processes themselves. Often this involves major capital investment and so such schemes have to compete with other projects that the firm has in mind in other, non-energy areas. Consequently progress in this area has been modest.

But how much of an impact can we make with energy conservation? Can it make a significant difference to our energy situation? The answer is that the impact could be enormous—in principle. In practice, there are many constraints, mainly of a nontechnical nature which we shall discuss later, which are likely to prevent an approach to anything like the full potential. There is an excellent report by Gerald Leach† which suggests that we could treble our gross national product, and hence our wealth, while still using only the same

† *A Low Energy Strategy for the UK* (International Institute for the Environment and Development, 1979).

amount of energy that we do now. The technologies he proposes exist, or more-or-less exist, and are more-or-less economic to deploy, but the energy savings which would be achieved in practice might not be as great as he suggests. However, I do not wish to indulge in a critique of his report. It is a valuable contribution to the debate on our energy future, and shows in principle what sort of savings we could achieve if energy conservation were our only national goal. Unfortunately, real life is more complicated, and we have many national goals to achieve simultaneously. This is another aspect of the social constraints that are likely to prevent us achieving in practice the full scope of the energy conservation potential.

Let us now look at some of these social constraints. There are two sides to the energy equation—supply and demand—and although they must obviously be related in that at the end of the day the equation must balance, the politics of the two sides are different. On the supply side, the government of the day either has direct control, as in the case of coal, gas and electricity, or has to deal with at most a few multinational companies, as in the case of oil. However, energy demand is in the hands of many millions of individual consumers. The government can exhort, cajole and even, through subsidies, offer bribes, but the ultimate decisions on energy use rest with the users themselves. So government action can have a direct and immediate effect on energy supply, but only an indirect and relatively slow effect on energy demand.

Only you can decide to improve the insulation of your house, change to a more economical car, or, if you run a factory, to invest in more energy-efficient equipment. It is difficult enough to get a government to make a decision. To get many millions of energy users to make their individual decisions—to get that many people to pull in the same direction at the same time—is obviously going to be even more difficult. So the hold-up is not the technology, or even the economics: it is social inertia and a lack of motivation. How do you get that many people to act?

There are things that the government can do to persuade

us to act in our best interests. They can, and have, mounted a massive publicity campaign to encourage us to 'Save It'. That at least has made us aware of the fact that perhaps we ought to be doing something. They could put up the price of fuel. The trouble with that is that it increases the cost of living, which finds its way into the next round of wage negotiations and we are back to square one. They could try legislation to require improvements in house insulation, as they have already. But that can only apply to new houses being built, and as the turnover of the national building stock is only about 2% a year, even when the building industry is booming, that is going to take a long time to produce any noticeable effect. They could, as they have also done, offer a subsidy to existing home owners to insulate their homes. It might seem odd to have to pay people to do things that are to their economic advantage anyway, but if it works then why not?

There is a further major social constraint to energy conservation and that is in cases where the owners of the property or energy-consuming device are not also the users. One example is the council-housing stock. The owners are the local authority and any investment decision to improve the insulation would be theirs and they would have to bear the cost, but it is the tenants of the property who would reap the benefit through lower heating bills. So what should the councils do? Should they insulate their housing stock at the expense of the ratepayers, which would cause them problems, or should they try to recover the money by increasing council-house rents, and attempt to persuade their tenants that they will still be better off because their lower fuel bills will compensate for the increase, which will cause them other problems. Local councils are not keen to make problems for themselves, and hence they have a certain reluctance to act.

Another notorious example is the company car. A large percentage of the motorists on our roads are not themselves paying for the fuel in their tanks. The firm is paying for it, and hence these motorists have no direct incentive to use it efficiently. You may have noticed that the swingeing increases in the price of petrol in recent years have had very little effect on the volume of traffic on the roads, or on the speed at which it travels. Perhaps now you can see why.

This divergence between ownership and use can provide a disincentive for energy conservation to even the best-intentioned people. I had occasion a few months ago to visit the offices of Friends of the Earth. They are a worldwide organisation with a mission, amongst other things, to persuade people to make wiser use of the Earth's resources—a worthwhile task they pursue with vigour and to considerable effect. The building in which they reside is owned by the Rowntree Trust, and along with the other occupants they pay no rent. The Trust also pays the heating bills for the building. Imagine my consternation on entering the offices to find that the heating was full on and the windows were open! I trust that this was but a temporary lapse from grace and that Friends of the Earth have returned to the fold of wise energy users.

Yet another social constraint on energy conservation is that energy isn't really all that expensive, even after the recent OPEC oil price rises. Certainly it is a lot more expensive than it was, but if you compare how much you actually spend on fuel with your other costs, such as your rent or mortgage, food, clothing, taxes, etc, you might be surprised to realise how small it is. In industry, although there are exceptions, fuel usually comprises only about 5% of total costs. That is very small indeed compared with wages, which can often represent as much as 60% or 70% of the total. It is not surprising, therefore, that a board of a company regards its Industrial Relations Manager as of greater value to the viability of the company than its Energy Conservation Manager, and consequently is prepared to give him a greater share of the company's resources to enable him to perform his duties.

Energy is the lifeblood of our industrial society, and it is the prospect of its nonavailability rather than its present price that is more worrying. Certainly many energy conservation technologies are economic, but the difficulty is persuading energy users that it is worthwhile making the effort when it represents such a small fraction of total turnover.

It is now some years since the oil crisis of 1973 brought the era of cheap energy to an end and convinced us to take energy conservation seriously. How well have we done so far? The answer is depressing. It is difficult to disentangle the effects of energy conservation from others such as the economic stagnation that the event also brought about. As nearly as we can estimate, in Britain we have achieved an energy saving of something like 6%, with an uncertainty of about 4%. As energy conservation is a field of diminishing returns, the first steps should be the easiest. That just is not good enough. We are going to have to do a lot better if we are to achieve anything like what we know to be the full potential of energy conservation, and achieve it before constraints on world supply really start to bite towards the end of the century.

As we saw, the main constraints are not technical or economic, but social. We need a radical change in our attitude towards energy. What we need is not new technologies, but the motivation to deploy them. There is a limit to what the government can do. After all, it is we, not they, who are using the energy. Certainly they can mount well publicised demonstration projects to show what can be done. They are doing a bit of that already—for example, they recently announced a 10-year programme aimed at reducing energy consumption by 18 million tons of coal equivalent per year—maybe they should do more. Certainly they could provide more advice on energy conservation. Why not local advice centres in every town—we have job centres and local tourist offices, for example, to help us in other areas—where local citizens could find out what they could do to reduce their fuel bills? But in the end it is you and I who must make the decisions that count—the decisions to act.

THE ROLE OF NUCLEAR POWER

I frequently find myself sharing a public platform with one or other of the leading lights of Friends of the Earth at meetings to discuss energy in general and nuclear power in particular. One of their number is fond of saying that he is quite prepared to accept that nuclear power is the answer, but what he would like to know, however, is: what was the question? A light-hearted quip, perhaps, but one that illuminates the essence of the problem. The question, of course, is: how do we meet our energy requirements in the future, particularly over the difficult period of the next few decades? It is a complex question, and one that requires, as we saw in the previous section, an understanding of the demand side as well as the supply side of the energy equation. It is far too complex a question, in fact, to permit simplistic solutions like 'let's build more nuclear power stations—end of problem'. Nuclear power is not *the* answer. At least it is not the whole answer. As an established technology, however, it could provide a major part of the answer.

It certainly is established technology, particularly in Britain. Development of civil nuclear power stretches back to the late 1940s. Progress, if slow, has been steady. Vast amounts of experience and expertise have been accumulated. The first nuclear power station in Britain, in fact the first in the world, was the Calder Hall station at Windscale (now Sellafield) in Cumbria which was switched on with great ceremony by the Queen in 1956.

HM The Queen opening the world's first nuclear power station at Cumbria in 1956. (© UKAEA.)

Calder Hall power station today. It is still working well after a quarter of a century. (© UKAEA.)

Britain had opted to go for the gas-cooled, graphite-moderated type of reactor, of which the early design, Magnox, formed the basis of the Calder Hall station. Following the success of Calder Hall, a number of Magnox reactors were ordered for various sites, and most of these became operational in the early 1960s. Taken together, these Magnox reactors produce some 10% or so of the country's electricity. Not a bad start for a relatively new technology, and one that augered well for the future. But the future was something that we did not give much thought to in the early sixties. We were in the era of cheap oil, after all, and the need for this new energy technology was not obvious at that time. So after a promising start, progress began to slow down.

The natural successor to Magnox was the AGR—the advanced gas-cooled reactor. A small 40 MW prototype had been built and operated successfully at Windscale (Sellafield), so the transition to this upgraded, but basically similar technology, should have been smooth. But what problems we had!

The 40 MW prototype advanced gas-cooled reactor (AGR) at Windscale (Sellafield). (© UKAEA.)

There were, of course, unforeseen technical problems but these were dwarfed into insignificance by the institutional and political problems. The sad and sorry story of the British AGR programme calls into question not so much our scientific and technical prowess, which in fact are pretty good, but rather our administrative and organisational skills. As a nation we seem to have lost the political will and organisational cohesion to tackle large projects successfully.

It is an instructive, if painful, experience to review the history of the AGR programme, albeit with the benefit of hindsight. The design and construction of the first Magnox stations (Calder Hall and Chapelcross) were administered by a single national body, the Atomic Energy Authority. The success of that programme persuaded the government of the day that nuclear power was now an established technology, and that the time had come to hand over its implementation to private industry. So they broke up the excellent design team at Risley—arguably the best in the world—into five mediocre design teams. Each of these was then attached to one of five consortia, made up of heavy-engineering firms with an interest in and capability for building nuclear reactor hardware. These consortia would then compete for orders for nuclear plant from the generating authorities both in this country, and hopefully overseas as well. At first things went fairly well. The consortia built the commercial Magnox programme without too much trouble. But when it was decided to order AGR for the next stage, each consortium was given one order for an AGR of 1250 MW capacity. By and large each developed a different design scaled up from the successful 40 MW prototype AGR at Windscale (Sellafield). It was a recipe for disaster and, not surprisingly, a disaster is what we got.

Scaling up by such a large factor proved to be too much for the decimated design teams to handle. One by one, the consortia collapsed and their work had to be taken over by those remaining. Eventually there was only one left, which had to do the lot. Effectively, one of the reactors—at Dungeness B—became the full-scale prototype on which the problems involved in such a large scaling-up were solved. Not surprisingly, Dungeness B fell behind schedule. Again not surprisingly, the whole programme fell behind schedule, and no export orders were received. After fifteen painful years we were back where we started—with one design authority. When the Chairman of the CEGB referred to the AGR programme as 'a disaster we cannot afford to repeat' it was to this appalling administrative mix-up he was referring, rather than the technology of the AGRs themselves.

What we *should* have done was to have kept the design team together, get them to build a full-scale commercial prototype based on their experience with the small Windscale (Sellafield) AGR, and then used that as the design for series ordering. The Canadians did just that with their own design, CANDU. They have been remarkably successful, not only in meeting their own domestic programme but also with export orders. CANDU is in no fundamental way better than AGR, but the Canadians have shown that they are better at organising themselves.

The AGRs are at last on stream, or are about to come on stream. It has been a long and painful experience, but if we learn from it, it will not have been in vain. We must learn to organise ourselves better in the future. And not just in our nuclear technology—our record with some of our coal-fired stations leaves a lot to be desired. Some have even taken as long to build as Dungeness B. When fully operational, the AGRs together with the existing Magnox stations will be supplying about 20% of our electricity.

Because of the problems with the AGR programme, there have been suggestions that we should change to some other design of reactor for future nuclear power stations. Following the oil crisis of 1973, the CEGB, who had previously been somewhat cool towards nuclear power, came to the conclusion that the changed situation demanded that they should go nuclear on a massive scale. They proposed placing an order for 23 new stations, and because of the deplorable state of the British nuclear industry they would have to go for the American-designed pressurised water reactor (PWR). The sheer magnitude of the proposed programme brought forth gasps of amazement from energy experts up and down the country. It had been based on a forward extrapolation of previous trends, which had become the common practice after two decades and more of steady economic growth. Electricity demand over that period had grown at about 8% per annum. Assuming that rate of growth was to continue and allowing for the fact that it takes about a decade to build a new power station, one unquestionably came up with the answer 23.

That assumption proved to be false, however, for the oil crisis of 1973 not only demonstrated the vulnerability of oil as a long-term energy source for all to see, it also triggered off an economic depression that is still with us. Rather than 8% per annum, the growth in electricity demand since 1973 has reached barely 2%. Furthermore, the power stations ordered in the previous decade were now coming on stream, which exacerbated the effect of the lower demand expectations and resulted in the Board having on its hands a surplus of generating capacity far in excess of any margin required to cover unexpected outages. So rather than 23 nuclear power stations, the CEGB decided that it really only needed two!

But there was still the question of reactor choice. At that

time the AGR programme was in a mess and nobody was inclined to order any more. So the choice lay between the PWR, which is what the CEGB preferred, and another reactor which had been developed in Britain in parallel with the gas-cooled reactors as a kind of insurance policy—the steam-generating heavy-water reactor (SGHWR). Work on the SGHWR has been going on at Winfrith in Dorset for many years, where a highly successful 100 MW prototype has been in operation since 1968. Claiming on this insurance policy as the need had arisen, the decision in favour of SGHWR was made early in 1974.

However, two years later the plans were back in the melting pot. The nuclear industry, following the AGR experience, was understandably nervous about scaling up from a small prototype to a full commercial size SGHWR; hence they looked at the problem in some detail. They came to the conclusion that the commercial launching costs of this new reactor type would be prohibitive, and advised the government to think again. By then, some of the AGRs had started to work and, based on the albeit limited experience of their operation, the CEGB were delighted with them. After fifteen years in the dog house, the AGR once again had its feet under the table. This reversal in its fortunes was hardly surprising—it is basically a good reactor, it was just the organisation of the construction programme that we messed up. So the order for two SGHWRs was cancelled, to be replaced by one for two AGRs.

Advocacy for the PWR is still strong, however. Indeed, the AGR and PWR are still under comparison in Britain, so it might be useful at this point to consider the relative merits of the two systems. In a previous chapter we noted that the thermal efficiency of the AGR is higher than that of the PWR—40% compared with 32%—and in this chapter we have noted how in this country we have acquired considerable experience, albeit painfully, in the construction of AGRs. To be set against this is the fact that the PWR system is the most widely used in the world and is likely to continue to be the preferred thermal reactor system in the future. There are over a hundred of them either operating or under construction in the world, and most countries with a nuclear power programme have chosen the PWR, whereas Britain's gas-cooled system has only been used in two overseas power stations—one in Italy and one in Japan. As there is a strong international component in the nuclear business, it could be argued that we should at least familiarise ourselves with the world's favoured reactor type. Also, so the argument goes, if we could establish ourselves in PWR construction we might be able to attract some of the export business, although we would, of course, be up against strong competition from other countries such as America, France and Germany who already have considerable experience in the construction and operation of such systems.

The safety characteristics of the two systems are different; detailed comparisons are hence difficult to make, but an attempt is made in a later chapter. Some experts believe that it is easier to make out the safety case for the AGR—for example, it operates at much lower pressures, has a single- rather than a two-phase coolant, has a graphite core which could act as a heat sink and has a prestressed concrete rather than a steel pressure vessel. On the other hand, because there are many more PWR stations in the world, much more experience of all aspects of operation has been accumulated than for the gas-cooled system. In order to meet British safety standards and to reduce radiation exposure of the operators, additional safety measures have to be designed into any PWR systems to be built in this country. This can be done and it is estimated that this will increase the capital cost of the plant by something like 10%.

Even with this extra cost it is estimated that a PWR will be up to 30% cheaper to build than an AGR of the same capacity, leading to lower electricity charges to the consumer. The major reason for the lower cost is that the PWR is largely factory built with the major components then being transported to the site, whereas a larger fraction of the AGR is built on site. This also has ramifications for the British reactor choice. If at least some of the major components for the PWR are built in foreign factories and then imported whereas the AGR is built on site using British labour, the implications in terms of balance of payments and domestic employment cannot be lightly ignored.

The argument could go on interminably and inconclusively, so something had to be done to resolve the issue. It has been decided that the next nuclear power station to be built in Britain will be based on the PWR system, provided it can meet the safety criteria of the CEGB and the Nuclear Installations Inspectorate and then emerge successfully from a Public Inquiry. On the basis of the experience gained, a realistic comparison can then be made with the AGR and a decision on which to use for future power stations can then be made.

After a long and sometimes traumatic development period, we now find ourselves with a nuclear capacity to supply about 20% of our electricity needs. Based on this experience we could, if we chose, expand our nuclear component. We have the technology and it is capable of implementation. Along with coal and energy conservation, therefore, nuclear power takes it place as one of the planks on which we can build our energy future in the difficult years ahead.

STATUS OF THE FAST REACTOR PROGRAMME

In the last section we discussed the thermal reactor programme, and saw that this was a fully developed technology and ready for implementation. What is the status of the fast reactor programme?

As we saw earlier, fast reactors can use uranium about 50 times more efficiently than thermal reactors, and so their successful development is a prize worth striving for. The development programme is being carried out at Dounreay on the northern coast of Scotland. A small 15 MW prototype, the Dounreay fast reactor (DFR) was built there in 1959, commissioned in 1962, and ran successfully until 1977 when it was closed down (it is now being turned into a museum). It was replaced by a much larger prototype the prototype fast reactor (PFR), with a capacity of 250 MW which started operation in 1976. Experience gained from operating the PFR to date has been extremely encouraging.

Dounreay on the northern coast of Scotland. In the foreground is the 250 MW prototype fast reactor (PFR). The sphere to the right is the 15 MW Dounreay fast reactor (DFR). In the background can be seen the outline of Orkney across the Pentland Firth. (© UKAEA.)

The next step will be to build a fast reactor of full commercial scale, the commercial demonstration fast reactor (CDFR), with a capacity of 1300 MW, to solve the problems involved in scaling up to full size before series orders for fast reactor power stations are placed. This is obviously a wise step, and should help us to avoid some of the problems we had when attempting to scale up the AGR. The CDFR is still on the drawing board, and once started is likely to take ten years to build. As with the PWR it will first of all require safety clearance by the Nuclear Installations Inspectorate, and will also be the subject of a Public Inquiry.

Fast reactors are likely to cost more to build than thermal reactors—up to one and a half times as much—but will be cheaper to run because they use so much less uranium. At the present price of uranium, a fast reactor could not produce electricity as cheaply as a thermal reactor of the same size, but it is expected that shortly after the year 2000 the price of uranium will have risen to a level at which this position will be reversed. So we would expect that by the late 1990s the fast reactor will be the preferred option for nuclear power station design, not least because it will help to stabilise uranium prices; hence we require our plans to be ready by then. Assuming that we make a start on CDFR within the next few years, we are just about on schedule to meet that date.

The development timescale of the fast reactor therefore dictates that it could be a major power source early in the next century. It is an option for the future and not, as are thermal reactors, an available technology to see us over the next few difficult decades.

ATOMS BY WIRE

The medium, or vector, through which energy from nuclear fission reaches the user is electricity. Hence it might be helpful at this stage to discuss the role that electricity plays, and is likely to play in the future, in the energy scene. Since it first became generally available at about the turn of the century, its use has grown dramatically. The rate of increase in demand has always been much greater than the rate of increase in demand for energy in general. During the giddy decades of the 1950s and 1960s electrical demand rose at about 8% per annum, whereas total energy demand rose by only about half that figure. Even over recent years, when overall energy demand has been fairly static, demand for electricity has risen by nearly 2% each year. So throughout this century electricity has been inexorably increasing its market share.

It is not difficult to see why this should have been so. It is a 'high-grade' form of energy, and can be put to a variety of uses. It is convenient to use and (at least for the user, if less so for those who live near the power station) it is very clean. It is rather more expensive than other forms of energy, but in the days when energy was cheap most of us were prepared to pay the extra premium for the convenience and flexibility it gave.

At a time when we have become more conscious of the need

to use energy efficiently, however, it has one grave disadvantage. The mechanism used to produce electricity in bulk is the steam cycle. A source of heat converts water into steam and this steam expands through turbines coupled to electrical generators. It is a fact of nature that when one converts energy from a low-grade form such as heat into a high-grade form such as electricity a large fraction of the initial energy goes to waste. This has nothing to do with engineering competence: it is a law of nature, known as the Second Law of Thermodynamics. In a typical power station, therefore, only about a third of the energy content of the fuel ends up as electricity leaving the station; the other two thirds end up as low-temperature heat discharged up the cooling towers or into the cooling water.

It often brings a smile to the faces of the old hands when newcomers to the energy field 'rediscover' the Second Law of Thermodynamics (one formulation was given by Lord Kelvin early in the nineteenth century). What an appalling waste of energy! Surely we must do all we can to avoid having to use electricity; let us use it only in those cases where there is no real substitute. As only about 8% of our energy demand is for applications that can be served only by electricity—such as lighting and electronic equipment—there is vast scope for such a change. However, the situation is not as simple as that, for two reasons.

The first is that although there is a large efficiency penalty to be paid in the production of electricity, once produced it can then be used with roughly 100% efficiency, whereas when raw fuel is delivered to the consumer's premises he must then use it in some device, be it a boiler or an engine, to convert it into useful energy and this conversion process itself carries an efficiency penalty. So what actually counts is the efficiency in end use, rather than at some point along the way from primary fuel to useful energy. Therefore in order to judge which is the most efficient method of using our precious fuels we have to examine each use on a case-by-case basis. In some cases, the electrical route is more efficient than the alternatives.

For example, let us look at rail transport. To drive a train by electricity, we have an efficiency of about 33% in first of all producing the electricity and the motor itself is about 90% efficient, giving an overall efficiency from primary fuel to useful motion of about 30%. On the other hand, a diesel engine has an efficiency of only about 25%. So it takes less primary fuel to drive a train by electricity than it does to drive it with diesel oil. To take a domestic example, if you have a full central-heating system fired by an oil or gas boiler, its efficiency is probably around 60–70%, which is far more efficient than using electricity. However, if you use that system during the summer months just to provide your hot water, because you have a large system feeding a small load the system must operate intermittently and hence its efficiency drops dramatically, probably to around 20% or so. If, therefore, you were to switch off your heating system for the summer and use an electric immersion heater instead to provide your hot water you will actually be using less of the nation's primary fuel. There are also many examples in industry where the electrical route is more efficient than the alternatives: drying processes, which are very energy intensive, are one such example.

The other reason why a judgment on the wisdom of using electricity is more complex than the Second Law of Thermodynamics suggests is that the fuels used to produce bulk electricity—low-grade coal, residual fuel oil and uranium—cannot readily be used for other purposes. The coal used is of such a low quality that it would not burn in a domestic grate, residual fuel oil is a thick tarry substance more like a solid than a liquid and certainly not usable in a domestic heating boiler, and what on earth would you do with a piece of uranium? In some countries, such as Norway, Switzerland and Canada, a large fraction of their electricity is produced from hydroelectric schemes. As this does not involve the steam cycle, the Second Law of Thermodynamics does not apply. In Britain, however, only about 2% of our electricity comes from hydroelectric power and there is little opportunity for expansion.

So it is in the nature of power stations to reject a large

fraction of the energy content of their input fuel. It is rejected at low temperature, in tepid water at less than blood heat, so the rejected energy is very low-grade energy. Nevertheless there is a lot of it. Can we not make some use of it? Indeed we can, although it is not all that easy to do so. The rejected heat is too low in temperature to be used directly for space heating, but we could build combined heat and power plants (CHP) in which the hot water is discharged at a more suitable temperature, such as 80 or 90 °C. This would reduce the efficiency of electricity produced even further, but as the rejected heat could now also be used for space heating, the overall efficiency of primary fuel use would rise to over 70%. There is unfortunately only limited scope in Britain for this technology because of our domestic lifestyle. Most of us prefer to live in individual houses, usually on estates surrounding our towns and cities. Installing the infrastructure of pipework necessary to transport heat to large numbers of such dwellings is prohibitively expensive. In some other countries, notably Sweden where the population preponderantly live in apartments, the situation is different. Supplying heat to apartment blocks from CHP schemes is much easier and much less expensive, and hence such schemes can and are used extensively in such countries. Nevertheless there is some scope in Britain, and the case for building a pilot CHP scheme on a commercial scale is a strong one.

Low-temperature rejected heat from normally operating power stations can be used to increase our indigenous food production. We talk of an energy crisis (when we really mean an oil crisis) now, but it will not be long before the world is talking about a food crisis—indeed some parts of the world already are. To take steps to produce more of our own food in Britain is obviously a worthwhile move. For example, rejected heat from power stations can be used for fish farming and for heating greenhouses. Indeed, there are a number of small pilot projects already in hand. It has also been suggested that we could use the rejected heat from power stations to dry grass for use as winter feed for our cattle.

So, because of the efficiency penalty involved in the gen-

Waste heat from a power station being used to heat a greenhouse. (© UKAEA.)

eration of electricity, we must look carefully at its use. In some cases an alternative method of energy supply will lead to a more efficient use of primary fuels; in some cases it will not. It is, however, a very flexible form of energy and is easily delivered to the user. It can be produced from a variety of fuels, many of which cannot be used sensibly in any other way. It is therefore likely that its market share will continue to increase. We saw how it was more efficient in terms of primary fuel to drive our trains with electricity than diesel oil. The electrification of our railways has already commenced, and is likely to continue as oil becomes less and less available. In the home, the cheapest fuel for heating at the moment is gas, but as supplies start to decline it is likely that the electric heat pump, a device which works on the same principle as a refrigerator and which uses electricity extremely efficiently, will start to penetrate this market. There are also many industrial processes in which the use of electricity can lead to increased fuel efficiency, and hence the use of these is likely to increase also.

With oil becoming scarcer, we are unlikely to want to continue using it for electricity generation as it has many other premium uses. Even residual fuel oil will have to be cracked down to produce other products as supplies start to dwindle. We will continue to use coal to make electricity, but again there will be pressure to use coal for other things—as a substitute for oil in many fields. It therefore seems highly probable that we shall need to expand our use of uranium, which has no commercial use other than electricity production, for this purpose. The technology exists and we have expended a lot of sweat in getting it right. We would be fools not to make use of it.

OTHER POSSIBLE SOURCES

Faced with a worsening situation over oil supplies, we have seen that we are going to have to take energy conservation more seriously, and to make more use of coal and also uranium. What other options do we have available? Are there any other sources of energy we can tap? What about nuclear fusion? What about the renewable sources of energy provided by nature arising principally from the flow of heat from the sun, and which man has used since time immemorial?

Nuclear fusion could provide us with vast amounts of energy if only we could master the technology. Fusion research in Britain is being carried out at the Culham Laboratory in Oxfordshire. At present we are building, in collaboration with our European partners, the Joint European Torus (JET). This is likely to be a ten-year project, and should bring us almost, but not quite, to the point at which we can demonstrate the principle of controlled thermonuclear fusion. On the basis of the experience gained from JET, it should be possible to build a super JET or, as the people at Culham will insist on calling it, jumbo JET, which will quite clearly demonstrate that we know how to control and sustain the fusion reaction in the laboratory. This also should be a ten-year project. Then we

The Joint European Torus (JET) under construction at Culham in Oxfordshire. Existing buildings of the UKAEA Culham Laboratory are shown in the background. (© UKAEA.)

must go on to build on this experience and construct a small prototype fusion reactor from which we can extract useful power—another 10 to 15 years—followed by a commercial-scale fusion power station on the basis of which we can start to order stations for use—again another 10 to 15 years' work. If we now add up the times for each step—and remember that each must be completed before the next one starts—then we see that this takes us somewhere towards the middle of the next century. So not until then, assuming that there are no major snags, can we reasonably expect fusion to be making a significant contribution to our energy supplies. It could well be a major energy source, but not for a while yet. Certainly it will not be with us in time to help us stave off the impending crisis towards the end of this century.

When we turn to renewable sources of energy the situation is frustrating. On the one hand, the total amount of energy available is enormous and inexhaustible. On the other hand, this energy is very diffuse and variable, being literally at the whim of nature. So actually making use of these sources on the scale we now use energy is not going to be easy. However, some progress has been made, and undoubtedly more will be. Let us review the present state of the art, subdividing these sources into the usual categories—solar, wind, waves, tides and geothermal energy.

The amount of energy reaching us from the Sun is truly enormous. In about three days we get about as much as our total energy demand for the year. The problem is tapping it. True, it normally appears as low-grade heat, but that does not matter because a large fraction of our energy demand is for this type of heat. In fact, some solar heat technologies are available—and in some situations economic—now: for example, in heating domestic water and swimming pools. Using it for space heating is more difficult. Because of its nature, the bulk of it arrives on sunny days in summer—just the time when we do not want heat to keep us warm in our houses, offices and factories. Solar panels will not be of much use to us on a cold February night, buried under an inch of snow.

Solar panels in use. They are already economic in some cases for domestic water heating. (© UKAEA.)

If only we could store it from summer to winter we could get round that obstacle, but that is not easy. The best we can be doing now is to design our buildings sensibly, so that we can take the maximum advantage of incidental solar heat gains and thereby reduce our demand for other fuels. Maybe a study of this aspect of solar energy should be an integral part of the training course for architects.

More grandiose schemes have been suggested, such as covering large areas of the Arizona desert with focusing mirrors to concentrate the Sun's rays sufficiently to raise steam to generate electricity. This would not be much good in Britain, however. Another suggestion is to build a large satellite in a stationary orbit with several square miles of solar cells to convert sunlight directly into electricity and then to beam the energy down to Earth as microwaves. Such a scheme is under consideration in the United States, but it will be many years, if ever, before it comes to fruition. But we have to consider projects on this scale if solar power is ever to make a significant contribution to man's energy supplies.

Another route we could follow is that adopted by nature. Plants are able to trap some of the solar energy falling upon them and convert it to chemical energy. Could we not therefore grow an 'energy crop'—known as 'biomass'—and either burn it as a fuel or ferment it to produce alcohol for use as a liquid fuel for transport? In Brazil they are in fact doing that. They are planning to produce enough alcohol from the fermentation of their excess sugar crop to make a significant contribution to their demand for portable fuel. Such crops require large areas of land, however; in a crowded island like Britain land is at a premium, and will become increasingly so as we need more and more land to grow our food. Even so, there are a number of interesting possibilities which are being investigated: for example, making use of agricultural waste or growing energy crops as catch crops or on marginal land.

Wind power is an indirect form of solar energy, produced by the interaction of the Sun's heat on the atmosphere. As we saw earlier, Britain had a wind power development programme in the late 1940s, which was run down when cheap

Middle Eastern oil started to flow. It actually got as far as building a 100 kW prototype on Orkney, which regrettably blew down in a storm. The Americans were also active in wind power technology in the 1940s, and had a 1250 kW device in operation for several years at Grandpa's Knob in Vermont. That, too, came to a sticky end when one of the blades flew off. Eight tons of metal hurtling through the air is scarcely conducive to public health. A few megawatts is about the largest practical size for a wind power device. It would be quite a large structure, with a blade span of about 150 feet, and over a thousand of them would be required to produce the output of one conventional power station. So we

Model of the 3 MW wind power device to be built on Orkney. (© UKAEA.)

are talking about tens of thousands of them to make a significant contribution to our energy demand. As they are such large structures, siting them might be a problem, particularly as the windiest sites tend to be on hill tops in areas of outstanding natural beauty. However, it is possible, though more expensive, to build batteries of windmills offshore in shallow coastal waters. This would increase the number of possible sites and would lessen the visual impact.

Wind power devices suffer from variability in their output, being dependent on wind speed, so the most likely way in which we would be able to use them is coupled into the National Grid as 'fuel savers'. When the wind blows, and output comes from the windmills, we can turn down the wick on our oil-fired stations and save precious fuel. Studies are in hand to investigate the problems of integrating such devices into our grid system.

Windmills are already marginally economic for isolated communities that cannot conveniently be connected to the National Grid and hence at present get their electricity from expensive diesel generators. In fact a 3 MW device is to be built on Orkney. As well as serving the local community it will also provide valuable experience in the construction and operation of such machines.

Wave power is also an indirect form of solar energy, produced by the interaction of the wind on the surface of the sea. It suffers much less from variability than does wind power itself because of the large smoothing effect provided by the motion of the sea. It is one of the renewable energy sources for which Britain is extremely fortunately placed, as we have such a long coastline per head of population. One favourable site exists on the coast of the Outer Hebrides where advantage can be taken of large waves rolling in across the Atlantic driven by the prevailing south-westerly winds. It is a site that is well away from the major shipping lanes, and there is no reason in principle why we should not place a string of wave power devices there of several hundred kilometres in length. That would certainly make a substantial contribution to the nation's electricity supplies. Furthermore, peak output would come in winter, which is the time when demand is greatest. Although it would be a massive undertaking it is susceptible to modular construction, with basic units of about 20 MW output which, once the design had been finalised, could be mass produced. They would be quite large units, however: about the size of a large ship. There is still much research and development work to be done before useful energy from wave power becomes a reality. At the present time, tenth-scale models are being tested on inland and coastal waters. This will have to be followed by sea trials under realistic conditions and then, if the trials are successful, the construction of full-scale prototypes. It is a programme, as with the development of nuclear fusion, that is likely to take several decades to complete. So although wave power stands out as one of the renewable energy sources that could well be of practical importance to Britain, the timescale of its development must mean that it cannot become so before the early part of the next century. It is a credible option for the future, but is

One of the techniques under development to extract energy from the waves—the Cockerell Raft—being tested in the Solent. (© UKAEA.)

unlikely to be with us in time to help solve our more immediate energy problems.

The rise and fall of the tides is caused by the gravitational attraction of the Moon on the oceans of the Earth. This regular motion of such a large body of water is another energy source which man can tap. In fact, one such scheme has been in operation at La Rance on the North Brittany coast for a number of years. Although, as with other renewable energy sources, its output is controlled by nature, that output is at least regular. We know that there will be a high tide every twelve and a half hours, and we can also predict with reasonable certainty how high each tide will be. Practical use of tidal power depends critically on the existence of suitable sites. Britain has an excellent site in the estuary of the River Severn,

A satellite photograph of the Severn estuary taken by the Space Department, RAE, Farnborough on 25 September 1979. The line to the right indicates the position of the tidal barrage suggested in the original proposal of 1933. The line in the centre is that suggested in the latest proposal. (© UKAEA.)

which is second in its suitability only to the Bay of Fundy in Canada. A quick look at a map illustrates why this should be so. The Severn estuary is like a gigantic funnel pointing out into the Atlantic Ocean. This concentrates the incoming tide, and amplitudes of up to forty feet sometimes occur.

Plans to build a tidal barrage across the Severn have a long history. The first such scheme was proposed in 1933. It was a scheme with quite modest output—about 600 MW—by modern standards, but not by standards of that time. It has been rumoured that had Hitler successfully invaded Britain in the 1940s, this is one of the schemes he would have put us to work on to take our mind off other things. The latest proposal is a more ambitious affair, involving a barrage from west of Weston-super-Mare to west of Cardiff, and a secondary basin to help make the output more controllable. It would have an output of between 2000 and 3000 MW, equivalent to that from about two conventional power stations. There would also be secondary benefits. The barrage could carry road and rail links across the Bristol Channel, and the harbour facilities in the port of Bristol would be improved.

Although dam building is a fairly well established technology, it is the size of the project that is daunting. It would be the largest civil engineering work ever undertaken by man, and apart from the Great Wall of China the only human artefact visible from space. Nevertheless, the scheme is under serious consideration and preliminary studies are underway. It would take up to twenty years to build, so once again this would be an energy option for the next century. If I might inject a note of political cynicism I feel that it is not our technical ability to undertake such a project that is the problem, but rather our political will to see it through. The length of time needed to build the Severn barrage is long compared with the lifetime of a government. What I fear might happen, therefore, is that one government will give the go-ahead, a few years later there will be a change of government and the new one in a fit of public expenditure cuts will then cancel the project. Such things have happened before in our recent history, and there is not much one can do with half a barrage.

It is not lack of technical skill, but an inability to organise and tackle large projects that is the most worrying feature of modern Britain.

Geothermal energy arises from heat coming from the Earth itself, partially due to the radioactive decay heat from naturally occurring radioactive substances in the Earth's crust, and partially due to heat conducted through the crust from the Earth's core. In some parts of the world, such as Iceland and New Zealand, where the geology is particularly favourable the heat actually comes to the surface as high-pressure steam. Under such circumstances geothermal energy is quite easy to use, and extensive use is made of it in these locations.

In other areas, known as 'semithermal' areas, useful heat

Geothermal energy. Steam geysers such as these make it relatively cheap and easy to exploit, but we do not have any in the UK. (© UKAEA.)

Drilling for geothermal heat in the Hampshire basin, next door to the Marchwood Power Station on the Solent. Water at 70 °C was found at a depth of 1600 m. (© UKAEA.)

can be extracted by drilling down a few thousand feet into the Earth. Sometimes in these areas aquifers occur from which hot water can be extracted at a sufficiently high temperature for use, say, for space heating. There is such an area on the outskirts of Paris, and use is made of it to heat a few thousand flats in the neighbouring area of the city. Britain has a few semithermal areas of this type: for example, there is one on the Hampshire basin not far from Southampton which is being investigated. Most semithermal areas are 'dry', however. This means that to exploit them two holes have to be drilled, the rock between them fractured, and water pumped down one hole and up the other. Again Britain has a few such areas and studies are being carried out on one of these in Cornwall.

So we can be hopeful that, in the long term at least, some of these renewable sources of energy can make a sizable contribution to our energy supply. Much work remains to be done on them, however, and so in the short term they are unlikely to be able to help us in any significant way. It would be most unwise, therefore, to include them in our present energy planning.

THE QUESTION OF ECONOMICS

Given the potential seriousness of our energy situation it might seem odd to be arguing about money. To a certain extent that is true, but nevertheless economics do enter into the argument. We saw, for example, when considering energy conservation how difficult it has proved to motivate energy users to take action even though it would be to their financial advantage. If it were not economic, there would be no chance at all. We saw also the economic problems that arose when the energy users, who would reap the benefit of reduced fuel bills, were different from the owners, who would have to lay out the capital investment necessary to achieve this saving.

Economics are also relevant to the supply side of the energy equation, particularly in the choice between different fuels. For example, it now costs about twice as much to heat one's home with oil as it does with gas. Not surprisingly, therefore, there is a strong preference for gas as a domestic heating fuel. More often, however, the discrepancy between alternative fuels is small, but nevertheless the subject of vigorous debate.

Let us look at the relative costs of producing electricity from the various fuels available. At the present time, nuclear power has a small but significant competitive edge over its rivals. Based on returns from the stations now operating, the present (1981) generating costs for electricity from nuclear power are about 1·65p per unit (kilowatt hour—that is, the electricity needed to run a one-bar electric fire for one hour), from coal about 1·85p and from oil about 2·62p. These costs include the capital cost of building the stations, amortised over a fixed period, and supplying them with their fuel. In the nuclear case it also includes an allowance for the eventual decommissioning of the station, likely to be more expensive than the decommissioning of a coal- or oil-fired station, and for disposal of the waste, which is likely to be more expensive than the disposal of waste from coal or oil. It does not include the costs of nuclear research and development, nor does it include the subsidies that are necessarily paid, and rightly so, to sustain the coal industry—sums which are, incidentally, comparable in magnitude. The price that we as consumers pay for electricity is about 4p per unit, the extra arising from transmission and distribution costs and overheads. So differences of a fraction of a penny compared with a market price of 4p are hardly world shattering. I am surprised that there is so much fuss about it. It has been estimated that with the fuel mix that we have at present for our electricity generation, the nuclear component, because it is marginally cheaper, produces a small reduction in electricity bills compared with what they would otherwise be. This works out at about £5 per year per household.

Although nuclear-generated electricity has an economic advantage over that generated from coal at present, it does not necessarily follow that we can extrapolate this advantage into the future. To understand why this is so, we must look at the differences in the make-up of the costs of producing electricity from these two sources. The major part of the nuclear generation costs comes from the capital cost of building the station; the cost of fuelling it, because it uses so little, makes only a minor contribution. With coal, it is the other way round. It costs much less to build a coal-fired station. It is the cost of supplying the fuel—about 5 million tons a year for a large modern station—that makes the major contribution to the price of coal-fired electricity. The capital cost is paid when the station is built. The fuel costs are paid continuously over the period of operation of the station. So at a time of inflation—and we have been suffering pretty badly from inflation of late—capital costs, invested some time ago, become less and less important compared with present running costs as each year goes by. It is inflation over recent years, therefore, that has been the main contributor to the present economic advantage of nuclear-generated electricity over coal. Looking at it positively, one could say that we took a wise decision in the past to invest our money in high capital, low running cost technology as a hedge against inflation. That, however, smacks of being wise after the event!

The comparative future costs of coal- and nuclear-powered electricity from stations being ordered now is not easy to predict. It depends on future price trends for the respective fuels and future construction costs. All the indications are, however, that nuclear power will continue to maintain its advantage. The best estimate is that from stations built in the near future nuclear electricity will cost about 2·5p per unit to generate compared with 3·5p per unit for coal.

The economics of the renewable energy sources are a greater cause for concern. It is estimated, for example, that the cost of electricity produced from the Severn barrage would be about twice as much as that produced from nuclear power. In the case of wave power, one of the renewable sources capable of expansion sufficient to supply a significant fraction of our electricity demand, the first estimate was that it might cost as much as 20–50p per unit. That is roughly ten times what we are paying now (Imagine your latest electricity bill multiplied by ten!). However, wave power development is still at an early stage, and we can hope that costs can be brought down as we learn more about this technology. Indeed, the latest cost estimates show that this is happening.

Another economic aspect we must consider is the amount of money we should presently be committing to the research and development of our various energy technologies. First of all, we must get this into perspective. Compared with commercial exploitation, research is extremely cheap. As a rough rule of thumb, for every pound we spend on research, we shall need about £10 for development and then about £100 to exploit the technology commercially. Research and development, then, should quite rightly be regarded as a non-recoverable overhead.

Economists have developed a technique, known in the trade as 'discounted cash flow', which relates the capital invested now in a technology to the value of future benefits to be gained from its use. Applying these standard rules of economics to research and development programmes will inevitably lead to nonsense answers. An amusing piece of nonsense of this type—at least in retrospect it was amusing—occurred in the 1960s. The government of the day, reviewing the fusion research programme, posed their economists the question: if we won't be using fusion power until the next century, how much should we be spending on research now? Not surprisingly the answer, from a mechanical application of the rules of discounted cash flow, came out to be nothing. The government therefore set about slashing the fusion research budget. The fact that if you spend nothing on fusion research it most certainly will not be ready for use in the next century, or any other century for that matter, seemed to have been overlooked. A final touch of irony is provided by the fact that this government had come to office under the slogan of 'forging the new Britain in the white hot heat of the technological revolution'. From where does one get the white hot heat if one insists on pouring cold water into the furnace, one is tempted to ask? Fortunately, the fusion research budget was restored to a reasonable level, but it took the oil crisis of 1973 to bring us finally to our senses.

There are therefore no foolproof economic criteria on which to judge the level of spending on research. It is a matter of judgment as how valuable one believes the technology will be to us when developed, how much can usefully be spent at any given time and what other competitive demands for research funds exist. In Britain at the present time (1981) we spend about £115 million a year on nuclear research and development, mostly on fast reactor development, about £100 million on the energy conservation technologies, mainly on the 10-year programme mentioned earlier, about £21 million on fusion research and about £15 million researching into the renewable energy sources. An immediate question that springs to mind is: are we spending enough on renewable energy research? Could these programmes move faster if they had greater funding? It is a basic fallacy, though a common one amongst those who have never engaged in research, to believe that the faster you pump money in the faster you will get results. Just occasionally a project reaches the stage where the availability of cash is the limiting factor, but this is not the normal state of affairs. Certainly it is not true for the renewable energy research programmes at their present state of development, although it might well become true in the future. Take wave power, for example. If that programme reaches the stage of building full-scale prototypes for realistic trials in the open sea, then indeed it will require funding on a much vaster scale than it does today. No doubt at that stage the necessary financial support will be made available.

THE NEXT FEW DECADES

We saw that in the long term there is no real shortage of energy. We do have a problem, however, with oil, and if we do not take action there could well be a critical situation developing towards the end of the century. What energy options do we have available to stave off the impending crisis? It is clear that the renewable energy sources, although of potential value later, are unlikely to be able to make a significant contribution by then. The same applies to nuclear fusion and also to fast reactor technology. It is a question of the timescale required to research, develop and implement these new technologies which precludes them from being ready in time. Even if someone were now to think of a bright new idea to solve the energy crisis, it is already too late. It should have been thought of twenty years ago—although at that time, with an abundance of cheap oil, nobody would have been interested.

We are left with just three weapons with which to fight our way out; these embrace established technologies for which we have just about enough time to deploy on a large enough scale. We have energy conservation, we have coal and we have nuclear power. We are going to need all three to see us through the difficult decades ahead.

Realistically, we ought to be able to achieve a 20% reduction in our energy demand, that is 20% less than it would otherwise be, by a sensible application of known and economically viable conservation techniques—although our record to date is not encouraging. We have previously discussed the Plan for Coal, designed to raise output from its present level of about 120 million tons a year to about 180 million tons by the end of the century. It is essential that we get somewhere near that target, for not only will coal be required to fulfil its present uses, it will by then have to substitute for oil and natural gas as well. Our present nuclear plant, including that near completion, is capable of supplying about 20% of our electricity needs. It should be possible to increase that to nearer 40% by the year 2000. That means ordering roughly one new station per year throughout the 1980s, a not insignificant but by no means impossible task.

That is the level of implementation of our three available technologies that we must achieve if we are to survive the impending crisis. Note that no one of them can see us through singlehanded: we are going to need all three. The next two decades are going to be difficult ones from the point of view of energy. The technologies exist. They are practical and economic. The targets set are not unrealistic. All we need is the political will to succeed.

CHAPTER 5

Matters of Public Concern

In the last chapter we saw how the lot of the common man has been improved dramatically by industrial development. It is also undoubtedly true that this industrial development has exacted its toll in environmental degradation. This has manifested itself in the dereliction of our landscape, the pollution of our air and our rivers, and in industrial accidents and disease affecting workers and in some cases members of the public as well.

'Where there's muck there's brass'. That was the slogan of the nineteenth-century industrialist and being as good as his word, he caused industrial society to mushroom exponentially amidst a flurry of smoke and slag heaps. As a result material living standards rose, but so did the amount of environmental pollution. Clearly there is a balance to be struck between the amount of wealth we want to create and the level of environmental impact we are prepared to tolerate as a consequence. Our own generation has become acutely aware of the effects of our wealth-creating activities on our environment—there is a limit to the amount of 'muck' we are prepared to put up with in order to make our 'brass'.

Energy is the lifeblood of industry, and the economic activity involved in the winning of the raw materials (coal, oil, gas and uranium), delivering them to the market place and converting them into useful forms of energy is a major industrial sector in its own right. As with industry in general, the energy sector also has its hazards and environmental impact. Coal mining is a long-established industry in Britain, and the dangers of working in the mines, either from pit accidents or from the long-term effects of inhaling coal dust, are well known. More recently we have seen environmental pollution on a grand scale from the foundering of large oil supertankers. There have been many personal tragedies resulting from the use of gas, either from its explosive properties or, in earlier days when town gas was used, from its poisonous properties as well.

The nuclear industry is relatively new when compared with our traditional energy industries, having been developed gradually in a number of countries since the end of the Second World War. It handles some extremely dangerous materials and is therefore potentially harmful to man. To date the safety record of the nuclear industry has been exceptionally good. At the very least that is a good start, and augers well for the future, but by no means let it be used as an excuse for complacency.

The various concerns that people have over nuclear power can be conveniently divided into two categories—the technological and the sociological. The technological concerns include the radioactive effluent from nuclear plant, a major release of radioactivity following some sort of accident, and the leakage of radioactive material from nuclear waste deposited under the ground or deep in the ocean. The sociological concerns relate either to the centralisation of the technology, or arise from the connection between the civil and military uses of nuclear energy.

The technological concerns can be summarised in one word—radiation. Many people have a deep-rooted fear of radiation. It cannot be detected by any of the senses and yet it can cause bodily harm the symptoms of which are not always immediately manifest. It has a sinister air of mystery about it. It is not a familiar or putatively understood hazard like fire, physical impact, chemical poison, or disease. How does radiation cause harm, and how serious a threat does it pose?

Before we attempt to answer this question let us digress for a moment to consider, in general terms, the biological processes that govern living organisms. Complex organisms such as man consist of many millions of individual cells. At a given time a small fraction of these cells is being damaged by things that the body comes into contact with in the environment. These include bacteria, naturally occurring chemicals in the air we breathe, the water we drink and the food we eat, physical impact with hard objects and, increasingly in more recent times, the effluent from man's industrial activity. Collectively these are known as 'biological insults' to the system.

Every living organism has developed a repair mechanism to replace cells damaged in this way. Following a major biological insult, however, the repair mechanism may not be able to cope and permanent injury or even death of the organism could result. Even minor insults are not absolutely free from harm. Due to imperfections in the repair mechanism they can, albeit with extremely low probability, provide sites for the subsequent development of cancer. If one of the damaged and subsequently imperfectly repaired cells is part of the reproductive system, this could lead to genetic defects in the next or even subsequent generations.

BIOLOGICAL EFFECTS OF RADIATION

Radiation can give up its energy to matter through which it is passing by the process of ionisation (see chapter 2) and this energy transfer can result in cell damage. So radiation is just one of the many biological insults to which living things, including man, are subject. What level of radiation dose would constitute a major biological insult, and what is the probability of low doses giving rise to cancer or genetic defects?

The basic unit of radiation dose is the rad. This is defined as an energy transfer of 100 ergs† per gram of tissue. The various types of radiation differ in their radiobiological effec-

† The erg is a fundamental unit of energy. It is very small, being equal to about 3% of a million millionth of a kilowatt hour.

tiveness: for example, alpha particles and neutrons are about ten times as effective as beta particles and gamma rays. In order to compare doses from different types of radiation, each is designated a 'quality factor' based on its radiobiological effectiveness, and a common unit of effective dose is defined as the dose in rads multiplied by the quality factor of the radiation. This unit is called the rem—rad equivalent in man. Recently, new units have been introduced internationally to replace the rad and the rem. These are the gray (1 Gy = 100 rad) and the sievert (1 Sv = 100 rem). However, in what follows we will stick with the old units as these are still more widely used.

A lot of information has been accumulated on the biological effects of large doses of radiation: from the medical records of those using x-ray sets in the early days, from laboratory experiments with animals and from the survivors of the atomic bomb attacks on Hiroshima and Nagasaki. One finds that doses in excess of about 1000 rem cause certain death within hours or days, at about 400 rem the probability of survival is about 50%; below 100 rem one will survive the acute effects but there will be a chance of subsequently contracting cancer. Below 10 rem it becomes increasingly difficult to find any statistically significant correlation between radiation dose and any deleterious health effects. This is because there are so many other ways of contracting cancer that these swamp the much smaller contribution one might expect from small doses of radiation. To check the correlation down to smaller and smaller doses one would need larger and larger populations of exposed individuals, and be faced with the problem of finding a control population of similar size which was identical in all respects with the exposed population except that they had not been exposed to radiation.

In the absence of direct evidence, how can one make some estimate of the effects of low levels of radiation? The effects at high levels are known and one can make the reasonable assumption that there is no effect at zero radiation. One then extrapolates down from the high levels to zero assuming that the effect is proportional to dose. This extrapolation is known as the 'linear hypothesis'. The vast majority of radiobiologists believe that this will give an overestimate of the effect at low dose, but it is felt to be wiser when dealing with a potential hazard to err on the conservative side. The linear hypothesis gives a correlation between dose and effect of about 125 induced cancers per rem per million people. That means that if you exposed each of a million people to one rem of radiation then about 125 of them would contract cancer at some time later in life. To put that number into context, about 200 000 of those million people will eventually die of cancer from some other, non-radiation cause. This illustrates the point that at low levels the adverse health effects of radiation, even on a pessimistic estimate, are very small indeed compared with the other hazards, natural or man-made, in man's environment.

A few years ago a study was made of the health records of employees at the Hanford plant in the USA. This is a nuclear establishment set up during the Manhattan Project during the Second World War, so these records cover a large number of people over a long period of time. The results showed that these employees appeared to have a higher incidence of cancer than is normal for the population as a whole. The effect was small but statistically significant. The implication of this result was that the effects of low levels of radiation were very much higher than suggested by the linear hypothesis. This is somewhat surprising, because this very much higher level of correlation leads to the conclusion that most naturally occurring cancers are produced by the natural radiation background (see later), which is obviously untrue. The explanation of this anomaly probably arises because of the fact that Hanford is in essence a chemical plant, and it is well known from other studies that workers in the chemical industry show a slightly higher incidence rate for cancer than the population as a whole. If one compares the Hanford records with those of chemical workers rather than with the general population, then the correlation between cancer and radiation disappears. The weight of radiobiological opinion is that the Hanford result is spurious, and that there is no reason to abandon using the linear hypothesis to provide a conservative upper limit on the health effects of low levels of radiation.

So far we have only considered what are called the 'somatic' effects of radiation—that is, the effect on the individual receiving the dose. We saw earlier that processes capable of causing cell damage can lead to genetic effects—that is, effects in the next or subsequent generations—if the damaged cells form part of the reproductive system. What then are the genetic effects of radiation? Extensive studies have been carried out on the Japanese bomb survivors, and one finds that those that received a dose in excess of 100 rem do show an enhanced somatic effect, but show no evidence of any genetic effects in their offspring due to radiation. Certainly it is true that some children born to the bomb survivors do exhibit genetic defects, but the magnitude is no greater than for the Japanese as a whole. So up to radiation levels which cause early death, in which case genetic effects are irrelevant, radiation does not produce any statistically significant genetic effect. That does not prove that there is no effect, only that it is small compared with that of the many other mechanisms that can cause defects in offspring, and that the genetic effects of radiation must be smaller than the somatic effect. The results of extensive experiments on animals enable us to put a conservative upper limit on the effect in man of about 80 defects in all subsequent generations per rem per million people exposed.

Radiation is not a man-made phenomenon nor the unique creation of the nuclear industry. Life on this planet has evolved in a mildly radioactive environment. The natural radiation background comes principally from four sources. The first is cosmic radiation—radiation coming at us from outer space, and deriving from the fundamental mechanisms which drive the universe. The second source is radiation from rocks, soil and building materials, which contain trace amounts of naturally occurring radioactive material and the third arises from radon, a radioactive gas which diffuses into the atmosphere from these materials. The fourth is from naturally occurring radioactive elements in our own bodies, chiefly potassium-40 in the blood stream. So each of us is a mildly radioactive source, slightly irradiating our fellows (and they us) as we come into contact with them day by day.

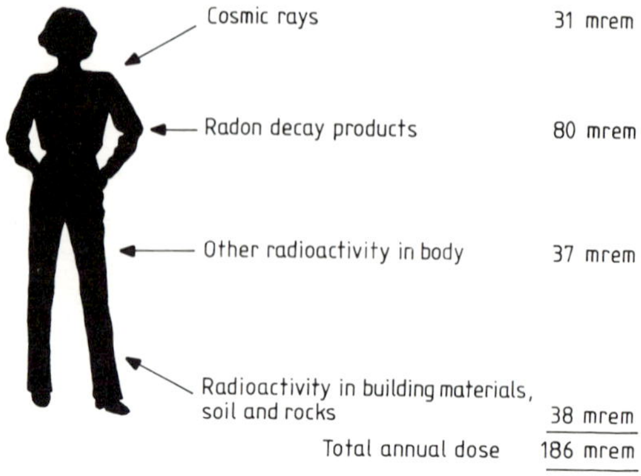

Natural sources of radiation. (Courtesy of UKAEA.)

The sources of natural radiation give each of us a total dose of the order of 186 mrem per year (a mrem—or millirem—is one thousandth of a rem). The natural background varies considerably from place to place. For example, it is about a third higher in Aberdeen—a city built largely of granite, a rock containing traces of uranium—than it is in London, which is built largely of brick. It is interesting to compare the rate of developing cancer in the two cities per head of population. Does the higher natural background in Aberdeen lead to a greater risk of cancer? On the contrary, the cancer rate is actually *less* in Aberdeen than it is in London, per head of population. This once again underlines the point that there are many mechanisms for inducing cancer, of which low levels of radiation are but one, and a very insignificant one at that. Fluctuations in the natural background of about a factor of two between different places in the world are not uncommon. Indeed, in some areas of Brazil and India the natural background is about twenty times that in Britain.

The process of evolution has equipped man with senses which, amongst other things, warn him of impending danger from hazards he might meet in his natural environment. For example, we can feel heat and see flame and thus avoid fire, and we can taste and smell some poisons. None of man's senses enable him to detect radiation. This leads one to suggest the hypothesis that Mother Nature does not regard radiation, at least at the levels man is likely to meet in his natural environment—say in the range of 100 mrem to 1 rem—as representing a significant hazard compared with the many others that he meets in nature. Our knowledge of the biological effects of radiation at these levels would tend to support such a hypothesis.

In the early days, when man first discovered and started to handle radioactivity and x-rays at around the turn of this century, little was known about the biological effects of radiation. It was not obvious that there were any effects at all because at the levels he was using it the effects were delayed and therefore not immediately apparent. There is a story told about Pierre Curie, one of the pioneers in the field, who strapped a piece of radium to his wrist for several weeks to see if it produced any noticeable effect. On removing it he discovered a radiation burn on the skin, but we shall never know whether or not this experiment would have produced any long-term effects on his health because, alas, a few days later he was run over and killed by a bus.

X-ray equipment was used extensively during the First World War as a diagnostic tool to examine injuries to war casualties, and in the years immediately following the war the rate of cancer deaths amongst operators of the equipment was observed to be rising alarmingly. Consequently, in 1928, an international body was set up—the International Commission on Radiological Protection (ICRP)—to examine the adverse health effects of radiation and to set limits on radiation exposure for people working with radiation. It initially set such a limit at 70 rem per year. Subsequently, on the basis of increased knowledge of the subject over the years, this was reduced to 15 rem per year in 1950, and finally to 5 rem per year in 1956. ICRP keeps the situation under constant review, but has not recommended any change to this level over the past twenty years. In practice, workers in the nuclear industry receive levels very much lower than this. Only a handful of them ever receive more than one rem. I have calculated that I receive more extra radiation from a week spent in Cornwall, an area with a much higher level of natural background radiation than in Oxfordshire, where I live, than I have accumulated in working at Harwell for ten years.

Although radiation cannot be detected by our senses, it can be detected very easily with fairly simple instruments. Each radiation worker carries his own personal dosimeter and the environment within which he works is checked continuously with highly sensitive instruments. He also receives regular medical checks. This level of precaution is unprecedented and unparalleled in industry in general. This has even led some of those of a suspicious frame of mind to suppose that it must be a very dangerous industry indeed if one has to take all these precautions. The suspicion should really be stated the other way round. Why don't other industries, many of which have much more dangerous working environments, take as much care as is taken in the nuclear industry?

The health and safety record of the nuclear industry is remarkably good. So much so that on any criterion it would be placed amongst the safest of human industrial activities. In almost all categories of disease and injury, workers in the nuclear industry compare favourably with workers in general. The only exception is in the category of road accidents, which shows a rate a few per cent higher than the national average (driving to all those public meetings I suppose).

Permissible radiation levels for members of the general public are set at a value ten times lower than that for radiation workers. There is a very good reason for this. Anyone who accepts employment in an industry or professional occupation also accepts, albeit tacitly, the risks that go with the job. Thus a coal miner puts himself at risk from mining accidents or from contracting pneumoconiosis, and a commercial traveller accepts an increased probability of being involved in a motoring accident. Although we seldom think about it as such, each of us accepts the risks associated with our chosen form of employment. It goes with the job and it is the job from which we earn our living. It is a 'voluntary' risk, even though we often accept it unconsciously.

Risks imposed on the general public from industrial or commercial activity, however, are 'involuntary' risks. There is no direct financial benefit to the public from the activity, nor is there any element of choice as to whether they wish to accept the risk or not. It is only reasonable, therefore, that risks imposed on members of the general public should be very much less than those imposed on workers in the industry creating the risk. The justification for imposing any public risk at all arises from the indirect benefit they get from the goods or services provided by that industry, as indicated by peoples' willingness to purchase such goods or services. Somehow this indirect benefit has to be balanced against this involuntary risk. Although there is no absolute and rigorous method for doing this, bodies have been set up in most industrial countries to recommend standards for the emission of potentially dangerous pollutants into our environment.

For the hazards of radiation, the appropriate body with this duty in the UK is the National Radiological Protection Board (NRPB), which bases its recommendations on those of the ICRP. The maximum permissible level set for radiation workers is 5 rem per year. The level set for members of the general public is ten times lower than this, i.e. 500 mrem per year.

The man-made source of radiation giving the largest dose to the general public arises from medical applications, such as x-rays and the use of radioactive isotopes for diagnostic or therapeutic purposes. In a sense this should be classed as a voluntary rather than an involuntary risk, because the patient derives direct benefit from this activity. For example, a single chest x-ray, which gives a dose of about 20 mrem, could well yield an early diagnosis of a potentially serious illness which can then be successfully treated, and this benefit must be balanced against the minimal risk of any possible harm from the radiation itself. The annual dose averaged over the population from medical applications such as these is about 50 mrem per person.

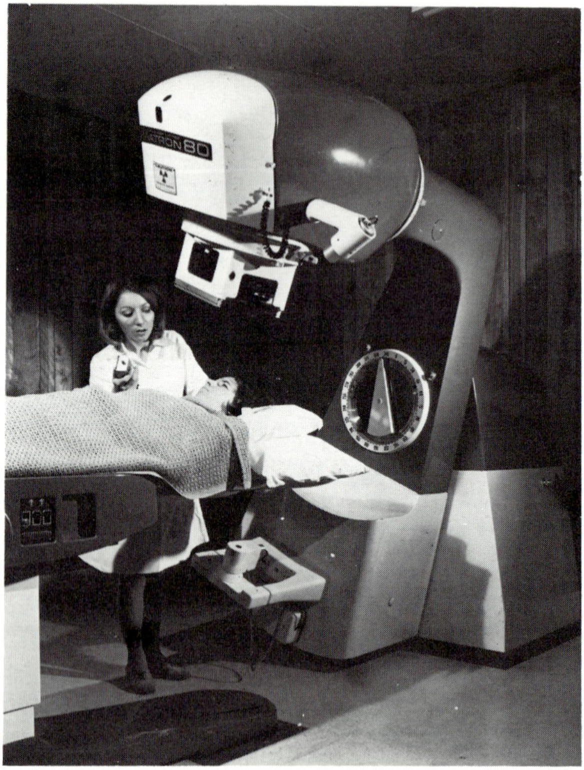

The medical use of x-rays has been of undoubted benefit to mankind. It also provides the largest dose of man-made radiation. (© UKAEA.)

Another source of man-made radiation giving some dose to the general public is from fallout due to the testing of nuclear weapons. Thankfully, since the signing of the Test Ban Treaty in 1963, which prohibits the atmospheric testing of nuclear weapons, this effect is declining and at the present time it contributes about 0·7 mrem to our average annual radiation dose. Some very ordinary things like television sets, which emit soft x-rays, and luminous clocks and watches, which have radioactive material incorporated in them to produce their luminosity, also make a small contribution. This is about 0·4 mrem per person per year.

From the activities of the nuclear industry, the average annual dose is about 0·2 mrem per person per year to a member of the general public in this country. Even with a very much expanded nuclear programme, it is not expected that this will rise to more than 0·5 mrem. That is half a mrem compared with a maximum permissible level of 500, and an average dose from natural background of about 100, and remember that this natural dose can vary by a factor of two or so depending on where you live.

Clearly no one can claim that such a small contribution poses a serious threat to public health. If it did then we would have to insist as a matter of urgency that, rather than close down the nuclear industry, we must order the immediate evacuation of areas such as Aberdeen and Cornwall—a course of action that would undoubtedly lead to vigorous public opposition!

But of course that is not the whole story. We have been talking in terms of average doses, but there will be individuals or groups of individuals who, because of their occupation, lifestyle or location, receive more than the average dose. Any radioactive emissions or effluent from a nuclear establishment must therefore be carefully monitored and controlled to make sure that nobody ever exceeds the maximum permissible dose. Part of this monitoring involves a detailed study of the biological pathways whereby radioactive effluent from a nuclear plant can find its way back to man. For example, a particular radioactive isotope might be concentrated in a certain type of plant which may then be eaten by animals, which may then themselves form part of man's diet. From such studies one can define 'critical groups' of individuals who might be most affected, and the level of effluent can then be set so that even these groups never exceed the maximum permissible dose.

Until a few years ago, the critical group most likely to be affected by the effluent from the Sellafield plant was the laver bread eaters of South Wales. For the benefit of non-Welsh readers I had better explain that laver bread is a green–black mush made from seaweed that a surprising number of people in South Wales (but, as far as I am aware, nowhere else in the country) fry up with bacon and eat. There is no accounting for taste, and the afficionados of South Wales actually regard it as a delicacy. One of the isotopes in the effluent from Sellafield is ruthenium-106. This isotope can concentrate in seaweed and thus find its way into the laver bread. The effluent levels from Sellafield therefore had to be set so that even a Welshman with an inordinate taste for laver bread would never exceed the maximum permissible dose.

A few years ago, the three people who used to collect the seaweed on the Cumbrian coast for the South Wales laver bread market finally retired, and so the laver bread eaters now get their seaweed from elsewhere and hence they are no longer the critical group. That distinction has now passed to the fishermen of Ravenglass, a small fishing port on the Cumbrian coast. The isotope caesium-137 can concentrate in fish, and so the effluent level from Sellafield is set such that no one in this critical group can possibly exceed the limit.

It is important when studying the various pathways whereby radioactive effluent can get back to man to do so in a quantitative manner. Otherwise spurious worries can be generated which have no relevance to the protection of public health. Two such red herrings emerged during the public inquiry into the proposed expansion of the Sellafield plant in 1977. It was suggested that effluent from Sellafield was contaminating the seaweed on the coast of the Isle of Man, and as this seaweed was used as a fertiliser on the island's potato crop, the public could be in danger by eating Isle of Man potatoes. Putting in the numbers, however, one found that an individual had to eat 30 tons of potatoes *a day* even to reach the maximum permissible dose!

Another suggestion was that tritium—a radioactive isotope of hydrogen—was accumulating in Lake Thirlmere, from which Manchester gets its water supply; hence the citizens of Manchester were being put at risk. Again, if we put the numbers in we find that such a citizen would have to drink ten thousand litres of water *a day* to reach the maximum permissible dose. This led the Inspector, Justice Parker, to remark: 'That will cause some sore bellies!'

So if you are in the habit of eating 30 tons of Isle of Man potatoes a day, washed down with ten thousand litres of Manchester tap water, then watch out! You could absorb enough radiation to give you a one in a million chance of contracting cancer later in life, to add to the about one in five chance of getting it anyway. There would, of course, be more obvious and immediate physiological effects of such a diet!

There are indeed hazards associated with nuclear effluent, but let's keep things in perspective. Although it would be arrogant—and also untrue—to claim that we know all that there is to know about the effects of radiation on man, certainly it is true to say that it is one of the best understood industrial hazards, and that it is controlled and monitored extensively. It will be a happy day for public health when other industries rise to this level of care over their effluent.

PROBABILITY AND CONSEQUENCES OF ACCIDENTS

What if something goes wrong? It is all very well to describe the minimal hazard to the public arising from a properly functioning nuclear industry. But suppose there is an accident—all human activities are prone to accidents—what price public safety then?

Accidents have indeed happened in the nuclear industry. In the early days of the American military programme there were some serious reactor accidents in which a few plant operators were killed.

The most serious nuclear incident to occur in Britain was in one of the military reactors at Windscale (Sellafield) in 1957. It was a very crude device compared with a modern civil reactor, consisting merely of a block of graphite moderator with holes drilled in it into which were placed rods of uranium metal canned in aluminium. Its purpose was to breed plutonium for the weapons programme, rather than to raise steam for electricity production, so the heat from the fission process was removed by a stream of air which was pumped through the holes in the block and then vented into the atmosphere through a tall stack. The graphite block was maintained at a fairly low temperature, therefore, and this gave rise to a rather peculiar phenomenon. The intense neutron bombardment affected the crystal structure of the graphite, causing the block to distort. This was clearly a nuisance, as it impeded the removal and replacement of the uranium rods. It was found that allowing the temperature of the graphite to rise slightly by reducing the air flow this distortion could be annealed out—a process known as 'Wigner energy release'.

The two early military reactors at Windscale (now called Sellafield) are on the extreme left and right of this picture. The one that caught fire in 1957 is the one on the left. (© UKAEA.)

This routine procedure was being followed one Friday afternoon—why do things always go wrong on a Friday afternoon?—when the temperature rose higher than expected, causing some of the uranium in one channel to melt and eventually catch fire. Smoke, laden with highly radioactive fission products, went up the stack. About 99% of this radioactivity was caught in the filter on top of the stack placed there in case of an incident such as this, but the remaining 1% or so escaped into the atmosphere. At first the wind direction was such as to take the radioactive plume out to sea, but a sudden change in direction brought it back inland, and radioactive fallout settled on the grass near the Windscale (Sellafield) site.

Milk being thrown away after the Windscale accident. The slogan on the lorry, although generally true, is perhaps inappropriate in this case. (© UKAEA.)

The most biologically significant isotope involved in the release was iodine-131. As some of this had fallen on grass which was being grazed by cows, there was a possibility that it could get into their milk, and if the milk were then drunk this could give rise to an unacceptably high radiation dose to the thyroid, particularly of young children, in which iodine accumulates. Iodine-131, however, has a comparatively short half-life of only eight days. So the procedure adopted was to ask the farmers involved to milk their cows as usual. It was then collected and thrown away. This regime continued for a period of a few months by which time the iodine-131 had decayed away. The farmers were compensated for their lost sales out of public funds, and rumour has it that the milk production from the farms affected rose dramatically over this period! There is no evidence to support the rumour, but it does alas accord with a fundamental aspect of human nature.

A thorough investigation was carried out shortly after the incident; this concluded that no member of the public nor any employee at the plant had actually suffered any harm. However, a more detailed study carried out recently by the NRPB suggests that, statistically, a few deaths may have resulted.

It will be recalled that the types of reactor used in the civil nuclear programme in Britain are also gas cooled and graphite moderated. However, they run at a much higher temperature than that of the ill fated Windscale (Sellafield) reactor; hence the distortion of the crystal lattice of the graphite and subsequent necessity for annealing are avoided. Also, they are cooled using carbon dioxide gas rather than air, so that particular type of accident could not happen to a commercial reactor. Other types of accidents have happened from time to time in the civil nuclear programme, and as any irregular incident involving radioactivity has to be reported no matter how trivial, and as these frequently appear in the press, the casual reader obtaining his information solely from the newspapers might be excused for thinking that all nuclear establishments are scenes of destruction and disaster. That this is in fact not the case can be readily verified by the reader visiting any of these establishments, each of which is eager and willing to organise such visits for members of the public.

The most serious accident to happen to a commercial reactor occurred at Three Mile Island power station near Harrisburg in Pennsylvania, USA, in March 1979. It all started in the non-nuclear part of the station. At 04.00 on 28 March the water supply to one of the steam generators failed. The turbines cut out as they were designed to do, and a reserve water supply should have automatically cut in. It did not do so because a valve which should have been open had in fact been left closed. It took eight minutes to correct this fault, and during that time heat was not being extracted from the reactor. So the reactor switched itself off as it was designed to do, and the pressure relief valve opened in its cooling system to allow excess steam to pass out into a dump tank. Unfortunately, this valve failed to close when the pressure had returned to normal and had to be isolated. While this was happening the pressure kept on falling, thus reducing the amount of coolant in the reactor. At this point the emergency core cooling system came into action, as it was designed to do. Then one of the operators, believing that the pressure vessel was already full of water, switched it off manually. Before this error was rectified, part of the core overheated and the zirconium alloy fuel canning interacted with the steam to form a bubble of hydrogen gas. Meanwhile, sufficient cooling water had passed over into the dump to cause it to overflow, allowing some of its vapour, containing a small amount of radioactive material, to escape into the atmosphere through the stack. Eventually, the emergency cooling was restored, the temperature of the core returned to normal, the hydrogen bubble dissolved, and the situation was gradually brought under control.

The reactor had been damaged extensively. It is estimated that it will take hundreds of millions of dollars and about four years of patient work to repair it. Repairing the damage to the public image of the nuclear industry might take a little longer, damage that was due mainly to confusing and conflicting statements issued from different official sources.

When news of the incident first broke, journalists from all over the world descended on Harrisburg like wasps descending on a honey pot. The prospect of reporting what could be the disaster of the century was an opportunity not to be missed. The fact that, as far as any harm to the public was concerned, nothing actually happened tended to lead to a sense of anti-climax. In fact the amount of radioactivity released during the incident was quite small. Even an individual who sat for the whole time of the incident at the power station boundary fence would have received a dose of only about 30 mrem. He could have got that much extra radiation by taking a fortnight's holiday in one of the areas of the world with high natural background.

Because some people had been conditioned into thinking that if anything went wrong with a nuclear reactor unprecedented death and destruction would inevitably follow, the fact that nothing actually happened has led them to refer to the Three Mile Island incident as a 'near miss'. It was not a near miss. It was a very serious accident indeed. The fact that the effect on the general public was minimal was not due to good luck; it was because nuclear power stations are designed with multiple containment systems to do just that. Some journalists still refer to the incident as a 'disaster'. One wonders what word they reserve for industrial accidents that kill people.

It has been estimated that the extra radiation received by the two million or so people living in the Harrisburg area as a result of the Three Mile Island accident might eventually lead to a 25% chance of one extra cancer death amongst them, that is one in the 400 000 or so that will eventually die from cancer from natural causes anyway.

As with all industrial accidents one can always say in retrospect: 'Oh, but it could have been much worse!' Indeed, it is the fear of a really big accident with a large number of casualties that worries people more than any number of much smaller but more probable accidents, even if the total number of casualties is the same. For example, I know from the statistics available that it is safer to travel a certain distance by air than by car, and yet I have a definite feeling of unease when sitting in a plane waiting for takeoff which I never experience when sitting behind the wheel of a car. An aircraft crash is dramatic and newsworthy and usually leads to a large number of casualties all at once. A car accident, on the other hand, usually only kills at most a few people, and would hardly get a mention in anything but the local press. Even so, far more people are killed on the roads in any one year than in aircraft accidents. The fact that you are just as dead if you die alone in a car crash as you are if you die along with hundreds of others in an aircraft crash shows how illogical the fear is. But illogical or not, it is still very real and we have to face up to it.

So, no matter how improbable it might be, we must ask which is the most serious nuclear accident that can happen and what the consequences might be. But to keep a sense of perspective we must compare this with the probability and consequences of other major accidents that could occur as a result of man's industrial activity.

One thing cannot happen to a nuclear reactor: it cannot explode like an atomic bomb. The reason for this is very simple. In order to get a nuclear explosion one needs more than a critical mass of fissile material at high concentration. Although there is more than a critical mass in the core of a reactor—otherwise it would not work—the concentration of the fissile material even in enriched nuclear fuel is only some 2–3%, which is nowhere near enough to cause an explosion. In some reactors, such as Magnox and CANDU, the fissile material is only at the natural concentration of 0·7%. It is substantially diluted with another non-fissile isotope of the same element—uranium-238—and there is no way that the concentration can be enhanced within the core of the reactor to anything even approaching that necessary to achieve a nuclear explosion. Furthermore, the fuel is arranged geometrically in the core of the reactor to be in its most reactive configuration. It has to be for the reactor to work. Any malfunction that disturbs this geometry must therefore render the core less reactive.

In the case of the fast reactor the situation is less clear cut. The fissile material in the fuel is plutonium with a concentration of up to 20% again diluted with uranium-238—i.e. the isotope of a chemically different element—and the core configuration is not in its most reactive arrangement. Even in this case, however, it can be shown that even under the most unfavourable condition the worst that can happen is a relatively small explosion—perhaps equal to that of a hundred pounds of TNT—the effect of which, although it might seriously damage the core itself, would not be felt outside the reactor. Thus a nuclear explosion is not one of the things one need fear from a reactor, and this is something with which even the most ardent critic of nuclear power would agree.

Having dismissed what cannot happen to a nuclear reactor, let us now discuss what can. The most serious thing that can happen is an event which is known as a 'meltdown'. The first thing that happens to a nuclear reactor if anything goes wrong is that it switches itself off. The control rods which are pulled out of the core to cause the nuclear reaction to start will drop back in under gravity and break the chain reaction at the first hint of trouble. Unfortunately, this does not then render the reactor completely safe. The reason is that heat is still being produced by the radioactive decay of fission products in the fuel elements. This heat output is about 4% of the full power output of the reactor one minute after shutdown, and about half a per cent or so after a day. So even after the reactor has been shut down, the cooling system for the core must be kept in operation.

Suppose the cooling system were to fail? There is, of course, an emergency cooling system that cuts in when the pressure on the primary system drops below a pre-set level, but suppose that were to fail also? There are various steps that the operating staff can take to intervene manually to provide cooling for the reactor core, but again let us suppose that for some reason they fail or are unable to do so. In other words, let us assume that, however unlikely, all means of cooling the reactor were to fail.

Under these conditions, the temperature of the reactor core would rise and eventually, after a period of several hours, the fuel would start to melt and would collect as a pool of very hot, very radioactive liquid at the bottom of the containment vessel. So far, although the damage to the reactor would be extensive—in fact, it would be a write-off—no harm would have been done to people because the dangerous material would still be held within the containment vessel. That would not be the end of the story however. The heat generated by the pool of molten fuel would be so intense that it is possible that it could, again over a period of several hours, burn its way through the base of the containment vessel into the bedrock underneath the reactor. There it would become diluted by molten debris, lose heat to its surroundings and eventually solidify.

On one occasion when this scenario was being discussed, a scientist commented loosely that the molten core would bore a hole through the Earth, coming out—if you happen to be standing in America, which he was—in China. Ever since this meltdown scenario has been nicknamed 'The China Syndrome'. He was grossly over-exaggerating, of course, but unfortunately some people took him seriously—even to the extent of making a film about it! In fact it would get nowhere near as far as China—a few tens of feet, maybe. But that is bad enough, for should the molten core breach the containment vessel the volatile radioactive elements in the fuel can escape into the atmosphere, giving unacceptably high radiation doses to personnel and members of the public in the area, up to a radius of possibly five to ten miles from the plant. Estimating the population distribution in the vicinity of a typical reactor site, and assuming that the authorities have taken no steps to evacuate them—again unrealistic, but we are trying to imagine the worst case—and assuming adverse weather conditions, one can calculate that the ultimate death toll (that is including all the long-term effects) from such a severe accident might be in the region of several thousands.

That is a major disaster in anyone's book, but unfortunately it is not the worst that man's technology can do to us. It is a brutal fact of our present way of life that one can construct disaster scenarios for many of our major industries—with large accidents having consequences far worse than that for a nuclear reactor. The reader might like to speculate—or rather brood upon—the worst that could happen following an aircraft accident (not just to those on board but also to those in the crowded city centre into which it might crash) the explosion of a multi-ton chlorine container (the lethal dose is a hundredth of a gram), or a major road accident in a heavily populated area involving a petrol or liquefied natural gas tanker, to name but a few potential nightmares. Not many industries have undertaken a study of the maximum disaster that they could inflict on the public. Of course many natural disasters, such as earthquakes or volcanic eruptions, can be much worse in their catastrophic potential than anything man can do, but it would be unfair to include these in any comparison because there is little that man can do to prevent them or mitigate the effects of consequences.

It is therefore necessary to assess a possible major nuclear accident in the context of our present industrial society. It is right that we should be apprehensive about such an accident, but not to the extent that it blinds us to the fact that larger and more probable accidents can occur in most other sectors of our economy. The remote possibility of major industrial accidents is endemic to our present industrial society. If we wish to take advantage of the benefits provided by such a society—and most of us, at least subconsciously, have decided that we do—then we have to be prepared to take the risks. I must apologise if I sound excessively morbid, but that is the way the world is.

The previous description of a major accident to a nuclear reactor indicates what might possibly happen in the worst case, but it is not certain that it will. Even after a complete loss of cooling water it could well be that the residual steam within the reactor pressure vessel might just about be able to provide sufficient cooling to prevent the fuel from melting, and there is some evidence from the Three Mile Island accident to suggest that this is so. Furthermore, if there were still a meltdown followed by a penetration of the containment vessel it could well be that the bedrock below the reactor might provide a secondary containment and thus prevent the emission of volatile radioisotopes. Again studies following the Three Mile Island accident suggest that this might be so. So it might turn out that even the accident scenario we described isn't credible after all, but until this is confirmed by further studies that would be an unwise assumption.

In our discussion of possible reactor accidents we have so far considered only the American-style PWR. As this is the preferred design in most countries with reactor programmes, that was a good place to start, but we must now ask to what extent does our discussion apply to the British gas-cooled reactors, Magnox and AGR. The American reactors use pressurised water as both coolant and moderator. Loss of pressure in the cooling system could lead to the water flashing to steam, which has significantly poorer cooling properties than water. The British reactors are cooled by high-pressure carbon dioxide gas and use graphite as a moderator. Loss of gas pressure leads to reduced cooling only in proportion to the pressure drop, a much smaller effect than that caused by a change of phase as is possible in a water-cooled system. A loss of coolant on a British reactor therefore means both that the gas pressure falls to atmospheric and that the circulating pumps stop working. Under these conditions the huge mass of graphite in the moderator acts as a large heat sink to absorb the radioactive decay heat emitted by the fuel elements. At the very least it would delay the melting of the fuel for several days, during which remedial action could be taken. Another difference between the two systems is that the PWR uses a steel pressure vessel whereas the British gas-cooled systems—with the exception of the early Magnox stations—use prestressed concrete. The significance of this difference is that should a significant crack appear in the pressure vessel it would spread very rapidly in steel but only slowly in concrete. Pressure vessel integrity is therefore more critical in a water-cooled system than in a gas-cooled one.

One could argue therefore that as a core meltdown is less likely to happen in a British gas-cooled reactor than in an American water-cooled one, they must be intrinsically safer. Reactor designers get very agitated when one puts forward such an argument. They prefer to say that it is easier to make the safety case for a gas-cooled than for a water-cooled reactor. This might sound like merely playing with words, but the distinction has some significance within the context of the British licensing procedures. The licensing authority (the Nuclear Installations Inspectorate) demands a required level of safety for any reactor proposed for construction in the UK. It might prove more difficult for an American-designed water-cooled reactor to reach that standard than for a British-designed gas-cooled reactor to do so, but unless it can it won't be built here and that's that. For example, the design of the PWR proposed for construction in Britain contains additional safety features over and above those incorporated in the standard American design to cater for accidents involving loss of coolant. A major study on the integrity of pressure vessels has been carried out and published, and sophisticated techniques have been developed for detecting very small cracks. These are important contributions to the quality assurance of the pressure vessel.

British safety standards and licensing procedures have a high international reputation. Most of the recommendations made by the Commission of Inquiry following the Three Mile Island accident would have the effect of bringing American procedures more into line with those in Britain.

Possible nuclear accidents can vary from the utterly trivial to a major disaster. The more serious the accident the less likely it is to happen. This spectrum of accidents is very similar to that which is found with other major technologies. For the longer established technologies one can draw a curve relating the seriousness and probability of an accident based on actual experience. It is interesting to note that if one draws these curves to the same scale then they all lie fairly close to each other. How would the corresponding spectrum for nuclear accidents look in comparison with these? That is a vitally important question to ask. As there have not been any serious nuclear accidents, in terms of harm to the public, one cannot draw such a curve based on actual events. The best one can do is to attempt to calculate it from the known failure rates of the components, including human beings, making up a reactor system.

A few years ago an American professor, Norman Rasmussen, conducted a major study in an attempt to do this. His report has been criticised, and as he admits rightly so, because the results are presented as a definitive curve relating the consequences and probability of a nuclear accident. Because this is a calculated curve rather than one based on actual experience the uncertainties in the final result must be much larger than for those for other technologies. Rather than a curve, therefore, his results should more realistically have been presented as a fuzzy band. However, be it a curve or a fuzzy band no one has challenged the basic conclusion that it lies significantly lower than the curves for other potentially hazardous technologies.

The general conclusion emerges clearly. Although one can have accidents at nuclear power stations and, albeit with very low probability, it is possible to have a very large one, the effects on ordinary people, both in consequence and probability, are very much less than those arising from many other activities that make up our industrial society.

WORRIES ABOUT WASTE DISPOSAL

Nuclear waste is very dangerous material. That statement is certainly true. Nuclear waste contains some radioactive substances with half-lives of tens of thousands of years. That statement is also true. It is not true, however, to combine those two statements to say that nuclear waste is dangerous for tens of thousands of years. Nor is it true to say that vast quantities of nuclear waste are piling up and no one has any idea what to do with it.

As we saw in a previous chapter, nuclear waste consists of about 40 radioactive elements and comes directly from the processes going on in a nuclear reactor. They can be divided into two categories. One is the fission products—which have half-lives up to about thirty years. They are principally beta and gamma emitters and so can potentially pose a radiation hazard to man from outside the body. The other is the actinides—formed by successive neutron capture in the heavy isotope of uranium—which typically have half-lives of thousands to tens of thousands of years. They are principally alpha emitters and so can only be harmful if taken into the body by swallowing or breathing.

We also saw that the waste will be converted into a vitrified solid encased in stainless steel with a corrosion-resistant alloy coating, and will be stored in this form in cooling ponds for several decades. If it were then ultimately disposed of a thousand feet down in granite, clay or salt, what potential harm might eventually arise to man and his environment from this practice, not just to the present generation, who have at least had the benefit of the electricity produced by its formation, but also to future generations who have not so benefited?

Nuclear waste has two advantages over other forms of toxic industrial waste. One is that the physical quantities to be handled are exceedingly small—only a handful per person if all his or her electricity for their entire lifetime were to be produced by nuclear power. This means that one can afford to take lavish care over its management without having any noticeable effect on the economics of the industry. The second advantage is that, unlike most other forms of dangerous waste, its hazard decreases with time.

It is interesting to compare the toxic potential of nuclear waste, that is the toxicity per gram times the number of grams produced each year, from a large nuclear programme with that of other dangerous materials routinely produced by industry in general and which are, regrettably, not handled with anything like the same care. Such a comparison reveals that nuclear waste is by no means uniquely hazardous, and is in fact far less so than substances such as barium and arsenic which are well known chemical poisons.

If one looks at the decay curve for nuclear waste over a long timescale one can clearly see the two components arising from the fission products and the actinides with half-lives of thirty years and about ten thousand years respectively. One also notes that after a few thousand years, that is when all the fission products have effectively decayed away, the waste is actually less hazardous than the ore from which it originated. So, contrary to a common superstition, it is the relatively shorter-lived fission products that are potentially harmful rather than the relatively longer-lived actinides. Nevertheless, a few thousand years is a very long time for most of us. That is why the waste is impounded in its glass and meal structure, which, tests have suggested, will maintain its integrity for at least as long as that, even if buried in unfavourable surroundings.

Although the long-term hazard is very small, it can be reduced even further by choosing a suitable geological formation in which to bury it. A site can be chosen with a long record of geological stability from which groundwater is absent. Granite, clay or salt deposits should provide suitable strata. Let me stress that one is not relying on long-term geological stability to provide radiological protection—which would, of course, be unwise as there is no way that one could guarantee such stability over millenia—one is merely making the rather obvious and sensible decision that as one has the choice one may as well choose the best geological site available.

If one were to have a major geological disturbance that were to bring up material from a thousand feet down that would constitute such a cataclysmic event that the fact that a small amount of that material was radioactive would be the least of our worries, and would make only a negligible contribution to the millions of casualties that would in any case ensue. More reasonably, one could ask what would be the consequences of the operation of Murphy's Law. Murphy's Law, as is well known to all engineers, states that if anything can possibly go wrong, it will. The application of that law to waste burial would be that even though one had chosen a good geological site, the day after it had been filled in some minor earth movement would cause cracks in the geological strata, even though they had remained intact for the previous many millions of years, such that groundwater would find its way into where the waste was buried. After about a thousand years the groundwater would start to slowly leach out the radioactive substances in the material. Let us further assume that by that time all records of the waste depository had been lost, and our future descendants decided to sink a well into this groundwater. What effect would it have upon them if they were to use it for drinking water? Even under the most pessimistic assumptions, the radiation dose they would get would only be a fraction of the natural radiation background. Under more probable assumptions, their dose would be several orders of magnitude less than that. As I said, the long-term effect of nuclear waste is utterly negligible.

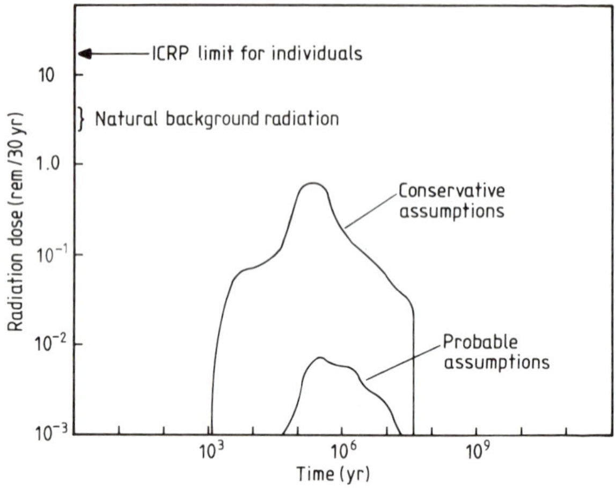

Long-term radiation dose to the public from a nuclear waste repository under both pessimistic and more probable assumptions. (Courtesy of UKAEA.)

One could describe the waste disposal strategy as a belt and braces—or even a double belt and double braces—technology in terms of its protection of public health. The quantities involved are so very small that one can afford to do that, so it is only responsible to do so. Other industries cannot possibly afford to take so much care over their toxic waste—the physical quantities involved are too large. This attitude, although technically sound and socially responsible, has produced an unfavourable public relations reaction. I have even heard it said that nuclear waste must be excessively dangerous if one has to go to such lengths to dispose of it.

It has always been a bit of a mystery to those working in the nuclear industry as to why it is that of all the potential environmental impacts of nuclear power, waste disposal should be singled out as the one about which the public are so obsessively concerned. It might have something to do with the fact that the waste management programme has such a long timescale, and because the programme is sometimes loosely referred to as the waste disposal 'problem'. The long timescale is determined, as we saw in an earlier chapter, by the half-lives of the radioisotopes involved—not, as some would have us believe, by the incompetence, or lack of knowledge of what to do, of the scientists and engineers involved in the programme. We must define what we mean by the word 'problem'. It is one of those 'anti-jargon' words that has one meaning in general conversation but quite another when used by scientists. In general conversation, the word is used to mean a difficult situation for which no one can suggest a solution. Thus the situation in Northern Ireland is a 'problem', and inflation is a 'problem'. The scientific definition of the word is a task to be performed—one can see clearly what to do but one has not yet completed it. It is in this sense, for example, that a tutor sets 'problems' for his students. There is no doubt (at least in the tutor's mind) that a solution exists, and the task to be performed by the students is to find it.

NO FREE LUNCHES

Some time ago, a friend of mine was driving along a country road near the town in which we both live when he came across a recently demised pheasant. Having Scottish connections, he got out of his car to take advantage of this lucky windfall. As he was doing so, a motorbike came round the bend, collided with him and broke his leg. Fortunately, no permanent damage was done and he has now fully recovered.

I quote this bizarre and apparently irrelevant story to illustrate a fundamental truth—there is no such thing as a free lunch. That is particularly true when applied to energy production, especially on the scale that we now use energy. There is no way that we can collectively produce and use the equivalent of about 330 million tons of coal each year, and that is just in the UK, without it having some effect on our environment and upon public health. We can and must take steps to reduce these effects as much as possible, but we can never eliminate them entirely.

We have gone to some lengths to discuss the hazards associated with the use of nuclear power. For the sake of completeness, and to demonstrate that these are by no means unique, let us consider for a moment the impact of other forms of energy production.

For example, let us take a look at our major fuel, and the one we have been using the longest time—coal. In the early years of this century the death toll from coal-mining accidents was horrendous. Over 300 miners lost their lives each year winning the coal that kept our society going. Much has been done to improve mine safety, and the toll is now down to about fifty per year.

The combustion of coal, and indeed all fossil fuels, leads to the production of a number of gases—sulphur dioxide, carbon monoxide and the oxides of nitrogen—which are known to be toxic. The lethal concentration levels of these gases are well known, and under normal conditions of operation their concentration never approaches anything like these levels, but what is the effect on public health of persistent doses at sub-lethal levels over many years? The disturbing fact is that nobody knows for sure, because it has not been studied in anything like enough detail. No one has yet established a dose against response relationship for these hazards in the way that it has been established for the hazard of radiation. At a rough guess, these poisons probably account for about two or three hundred deaths per year amongst the general public.

There is even a small amount of radioactive effluent coming from a coal-fired power station, arising from the trace amounts of uranium and thorium—at about one part per million—that occur in coal. In fact the radiation dose you would get living near a coal-fired power station is about the same as that which you would get living near a nuclear power station. Both are utterly trivial, however, in terms of their deleterious effect on public health.

Perhaps the most worrying long-term effect arising from

the combustion of coal—and indeed that of any fossil fuel—is due to the emission of carbon dioxide. This gas is a direct combustion product and is produced in large quantities; nothing can be done with it apart from discharging it into the atmosphere. It is a non-poisonous gas, but in sufficient quantities the fear is that it could produce a disastrous effect on the climate.

TABLE Discharge rates of some pollutants from a modern 2000 MW coal-fired power station.

Discharge	Tons per day
Fly ash	20
Sulphur dioxide	500
Carbon dioxide	50 000
Carbon monoxide	4
Nitrogen oxide	160
Radioactivity (thorium and radium) 0·15 mCi per day	

The Earth as a whole is in a state of thermal balance. It absorbs heat from the Sun—which comes in at optical wavelengths—and re-emits it back into space at longer wavelengths, known as infrared. The amount absorbed as light equals the amount re-emitted as infrared, thus maintaining the thermal balance which maintains the Earth at its present temperature. Carbon dioxide has the property that it is transparent to light but is opaque to infrared. If the concentration of carbon dioxide in the atmosphere were to rise too high, therefore, the same amount of energy would come in from the Sun but less would escape into space. This would cause the average temperature of the Earth to rise. The balance is a delicate one. It would only take a rise of a few degrees to melt the polar ice caps, with consequences that hardly bear thinking about. A large proportion of the land area of the globe would disappear below sea level. This property of carbon dioxide is rather similar to that of glass. This also transmits light but absorbs heat, and this property is used to advantage in horticulture to increase the environmental temperature for the benefit of certain crops. Hence the possible large-scale effect of carbon dioxide on our climate is known as 'the greenhouse effect'—an appropriate descriptive title.

Nobody knows at what level of concentration of carbon dioxide in the atmosphere this catastrophic and irreversible change will take place. There is no doubt that the level has increased measurably since the time before the industrial revolution, and it is still rising. The bulk of this rise is undoubtedly due to man's combustion of fossil fuels on a massive and increasing scale. There is also another of man's activities which is compounding the problem. Carbon dioxide occurs naturally in the atmosphere, of course, and is part of the cycle of life. Animals, including man, breathe in oxygen and expire carbon dioxide. Plant life absorbs carbon dioxide and emits oxygen, and so the balance of carbon dioxide is maintained in the atmosphere. The huge forests that occur in various parts of the world are a large part of the plant life maintaining the balance—forests which man is systematically destroying! So perhaps chopping down trees, thus preventing them from absorbing carbon dioxide, and then burning them in a wood stove, thereby emitting more carbon dioxide into the atmosphere, is not such an environmentally sound practice after all. An acceptable practice for just a few people, perhaps, but disastrous if all the four billion people on Earth were to do it.

So far we have only considered the slow and the cumulative effect on the environment through the use of fossil fuels. Just occasionally the impact has reached disastrous proportions. There was the case of the London smog of 1952. Smog arises when the combustion products of coal become incorporated in fog. In the winter of 1952 the climatic conditions were just right, or from the point of view of the inhabitants of London all wrong, to produce a pea-souper to end all pea-soupers. London had experienced smog before, but not on a scale such as this. For many days that winter, Londoners coughed and choked themselves to a standstill. There had been nothing like it in living memory, and it all arose from an activity vital to the wellbeing of industrial man—the burning of coal. The old and those with chest complaints were particularly badly hit. It has been estimated that the London smog of 1952 was responsible for about 3900 premature deaths.

There was a public outcry, as indeed there should have been. As though by way of consolation, at least some little good came out of the incident. Parliament passed the Clean Air Act, which controlled the type and amount of fuels that could be burned in built-up areas. It has been argued that this is the only sensible piece of legislation passed by Parliament in its long history! A bit of an exaggeration perhaps, but certainly this Act must count amongst the better pieces of work Parliament has done for us. It is a pity that it needed such a tragedy to spur it to such heights.

Some years later there was another tragedy of somewhat different kind. Although the toll in human life was smaller, such was the nature of the disaster that it stunned the nation. Even today, many years later, the scars have still not healed; people are stilled into silence by the mere mention of the place where it happened—Aberfan. Aberfan is a small village in the coal-mining district of South Wales. Coal had been mined in the area for almost a century, and huge tips of colliery waste had accumulated outside the village over that time. At 09.30 on 21 October 1966 one of the tips slipped and thousands of tons of mud and slurry came washing down on the village school. Over a hundred children were killed. 'And the sins of the father will be visited on the children even into the third and fourth generation.' If only our forefathers had been a little more careful, in their pursuit of wealth, with what they had done with the waste from their operations. It is a lesson that we must not fail to learn.

In 1979, an oil tanker was unloading at the terminal in Bantry Bay in Ireland. Suddenly there was an explosion, followed by a giant fireball. Fifty men died instantly. It could have been a lot worse if the fire had spread to the storage tanks on shore, which, fortunately, it did not. Even hydroelectric power has not been without its casualties. Within the last eighty years there have been two disasters, one in the

USA in the 1890s and the other in Italy in the 1950s, with death tolls in excess of two thousand.

We could go on and list many more disasters, some large and many small, but to do so would be to lapse into excessive morbidity. In any case it is unnecessary, for the point has surely been made. This is the price in terms of human suffering that we have had to pay to supply energy to our industrial society. It might come as a shock to realise that it is a price that society has agreed, albeit tacitly, to pay, for we all continue to use energy, and in ever increasing quantities, and we continue to demand the goods that our energy-based industrial society can provide. There is a lesson to be learned from all this. It is one of the features of a sustainable civilised society that it can learn from the mistakes of the past. When developing a new technology, and (within our context) specifically a new energy technology, we must make sure that alongside the technical development we also develop an understanding of its potential hazards and take steps to minimise them. It is another feature of a sustainable civilised society that it does not sit down in a corner holding its head in woe when faced with challenging problems, but rather sets its mind to solving them. A society that yields to defeatism has no right to survive.

I trust that it has by now become clear that it is energy production in general, and not just the portion produced by nuclear power in particular, that is potentially harmful to man and his environment. This applies to persistent low-level effects, cumulative effects and major disasters alike. Indeed, we must continue to control and monitor the activities of our nuclear industry. We must continue to learn more about its potential hazards, and take all necessary steps to minimise them. Equally, we must apply the same vigilence to our other energy industries.

It is, of course, no defence of the nuclear industry to say: 'if you think that we are bad, take a look at them!' Such a comparison does have value, however, because if we are to introduce a relatively new technology and to implement it on a large scale, it is reasonable to ask that it be at least as safe and its environmental impact be at least as small as those technologies which it is replacing or supplementing. It should preferably be a lot safer and its impact considerably less.

In order to make such a comparison fairly, it is important to be sure that we are comparing like with like. We must assess the full fuel cycle in each case, from the winning of the basic raw material to the ultimate disposal of the waste products, and we must compare the relative impact of each for the same amount of energy produced. A number of such comparative studies have been performed in several countries and they all tell the same story. To produce a given amount of energy from nuclear power will cause less harm than producing it from coal or oil. The same conclusion is reached whether one considers harm to employees in the respective industries or to the public at large.

What about the new and as yet undeveloped technologies such as nuclear fusion, or the natural energy sources such as the Sun, wind, waves, tides and geothermal energy? Even they will have some environmental impact and could potentially cause harm to people. For example, fusion involves handling the gas tritium, which is a radioactive form of hydrogen. The shield surrounding the plasma will suffer from intense neutron bombardment and will have to be changed frequently, possibly each year, and that will provide a large quantity of highly radioactive debris to be disposed of. The natural energy sources are by their nature very diffuse, and so to collect the amount of energy equivalent to that produced by a nuclear power station large structures, or equivalently a large number of small structures, will have to be manufactured. Many tons of steel, glass and concrete, etc, will have to be produced, and one knows from bitter experience the harm that would follow from the production of these.

Intuitively one might feel that on the basis of present knowledge the overall impact of these technologies would be less than from nuclear power for the same output of energy, although the only in-depth comparative study that has been performed to date does suggest otherwise. But then it is in the nature of man to believe that the grass is always greener on the other side of the fence. As I said—there is no such thing as a free lunch!

THE TECHNOLOGICAL IMPERATIVE

So far, we have considered only the technical impacts of nuclear power on man and his environment. To some extent they are quantifiable and can be put into the context of our industrial society. In particular, they can be compared with the impacts of alternative methods of generating power using our more traditional fuels. Comparing like with like, we found that for an equal amount of power produced, nuclear power was no worse in its impact, in fact it was considerably better, than most other fuels.

We must now go on and consider the social implications of using nuclear technology. By their nature, such impacts are much less easy to quantify, and so it is less easy to assess them in such a definitive manner. They are nonetheless real for that, and they constitute a very genuine, and perhaps the most justifiable, source of public concern.

The introduction of a major technology, and its implementation on a large scale, is bound to lead to some change in the structure of the society into which it is introduced. This phenomenon has been observed many times throughout history, from the invention of the wheel to the mastery of powered flight. It has even been given a label—the technological imperative.

Let us take one example, the invention of gunpowder and its introduction into Western European society. Prior to this, mediaeval Europe was basically a feudal society, with effective power residing with the local barons. True, they professed some sort of loyalty to an overlord, usually the one of their number with the strongest personality, but it was a loyalty of consent and convenience and consequently variable depending upon the circumstances prevailing at the time. The overlord had no way of enforcing his sovereignty, because when under threat the individual barons could retire to impregnable fortresses they had built on their own lands. With the invention of gunpowder all that began to change. The overlord now had the power to knock down the castles of his barons, or if he did not have the strength of will to do so he was supplanted by somebody who did, and so establish himself as an absolute monarch. The era of the sovereign state within well defined geographical frontiers had begun. It is an era that has persisted down to the present day, when the sovereign states of Western Europe are taking the first faltering steps towards collective action to preserve their independence and influence in a world dominated by superpowers—the technological imperative brought about by another invention, the atomic bomb.

The introduction of nuclear weapons has undoubtedly had a dramatic effect on the social and political structure of the world. The world suddenly became a different place on 6 August 1945 when the full extent of the power of these weapons was first demonstrated over the unfortunate Japanese city of Hiroshima. The question we must ask, however, is what are the social effects of the use on a large scale of nuclear energy for civil purposes? Were it not for the fact that civil nuclear energy can possibly be abused for non-peaceful purposes, its technological imperative would be relatively small. The major question, therefore, is to what extent the use of nuclear energy by responsible people for good will assist its use by irresponsible people, be they terrorist groups or governments, for evil. Or, phrasing it the other way round, if responsible people forswore its use for good would that in any way prevent its use by irresponsible people for evil?

THE CENTRALISATION OF POWER

Before considering in detail the various ramifications of that major question, there is another aspect which we must discuss. By its nature, nuclear technology is not well suited to direct use in the home or even in the local community. On the contrary, nuclear technology is big technology, with large unit sizes handled by large institutions, and large institutions mean large, centrally controlled bureaucracies. There is growing public unease, not just in Britain but in most of the developed democracies, over the growth of large centrally controlled bureaucracies and the increasing domination they are exercising over our everyday lives. There is a strong counter-movement demanding a decentralisation and devolution of power to levels nearer to the local community, a breaking up of the large institutions into smaller units more compatible with the human scale. There is a feeling that our society is overcentralised, to the extent that the ordinary individual citizen no longer matters and that he is no longer able to influence or control the decisions that affect him. Enough is enough. Let us re-assert and re-establish the human dimension in the conduct of our affairs. It is a movement aptly summarised by the phrase 'small is beautiful'.

It is a movement with which I have the strongest sympathy and which I actively support. There is much in our society that is unnecessarily overcentralised and overbureaucratic, and it is high time that we set about restructuring our society to put these things right. For example, is it really necessary for Parliament to waste its time debating legislation which controls what the citizens of Abingdon can and cannot do with their own market place? Or that the Secretary of State for the Environment should himself have to make a decision as to whether a local parish council should or should not enforce a preservation order on a local tree? And there is that vast bureaucratic load of interlocking nonsense the Regional and County Structure Plans—a master plan of Gargantuan proportions that will prove to be a millstone around the necks of the present and future generations, and which should have been strangled at birth!

However, one must beware of making a fetish of decentralisation. We do live in a complex interdependent society, and some central control is essential to enable so many of us to maintain our present standard of living on these small islands. For example, for as long as it is necessary to defend ourselves from the threat of external attack, I can see no advantage in decentralising the armed forces. Nor do I think there is much to be gained by each local council conducting its own independent foreign policy. There are also many essential services which are best organised nationally, such as post and telecommunications, and the railway system. As with all things, it is a question of balance. Some activities are best organised locally, others regionally and yet others nationally. The object is to get the scale right for each. We might indeed in the recent past have leaned too far towards centralisation. We will get it wrong again if we now react by swinging too far the other way. Small might be beautiful, but the right size is best of all.

How does this apply to that sector which nuclear power can serve, the supply of electricity to individual users? At present, the generation and supply is by and large a nationally organised service, although there is still a small amount of private generation by large firms, and the national organisation does have a regional structure. It should be pointed out that this style of organisation has nothing to do with nuclear power, and was already established long before nuclear power was used to generate electricity. When electricity first began to be used at about the turn of the century, it was supplied locally by small private firms. Gradually, as its use became more widespread, amalgamations took place leading to somewhat larger organisations but still on a local scale and usually run as a local authority undertaking. Eventually, after the Second World War, these were all combined into a national system. The detailed structure of the organisation may have changed from time to time since then, but the principle of a National Grid fed by central power stations was firmly established.

The system has two major advantages. The first is that power can be readily transported from one part of the country to another via the National Grid. So local supply does not depend on the functioning or otherwise of the local power station. Thus continuity of supply is more easily achievable for all users. The other advantage is that of cost. Power stations can be built as large units, as they are putting their output into a National Grid rather than just meeting local demand. Thus advantage can be taken of the economies of scale, which apply with some force to power station construction, at least up to the present size, although there is some argument as to whether this will continue to apply for even larger sizes.

Most homes and other establishments in the UK are now connected to the National Grid, although this is not universally so. There are some offshore islands, for example, that are not so connected, and which have to rely for their electricity on small local diesel sets. Unfortunately, apart from problems when the set breaks down, they find that they are paying about five times as much per unit as we are on the mainland. They are finding that although small might be beautiful, it can be very expensive! On balance, and remember that many of these islanders are true Scotsmen, they would prefer to be connected to the National Grid.

There are, of course, some disadvantages with the present system. The power stations are controlled by at most a few hundred operating staff. If they decide to go on strike, the whole country is blacked out. However, even if we had a collection of smaller local operations, the staff would still presumably all belong to the same union, so the effect of a strike would be rather similar. So to avoid the problem one would not only have to decentralise the electricity supply system, one would have to decentralise the unions as well!

Another problem with large centralised power stations is that it is difficult to make effective use of the waste heat they produce. As we saw in an earlier chapter, only about a third of the energy content of the fuel ends up as electricity, and the other two thirds is rejected as low-temperature heat. It is relatively easy to transport electricity over long distances, but not quite so easy or economic to move low-temperature heat in bulk, and there is not much that can be done to use all that heat locally to the station. There is a strong case for refurbishing old, and usually smaller, stations which tend to be near centres of population and industry as combined heat and power units to supply heat to local users and power to the National Grid.

Thus large nuclear power stations can be fitted naturally into an existing organisational structure—a structure that was already there before they appeared on the scene and which they were in no way responsible for shaping. Nuclear power stations may be large centralised units, but so are coal- or oil-fired power stations. So long as there is a demand for bulk supplies of base-load electricity, then nuclear power stations have a role in helping to supply that need. A large centralised electricity supply system is not a technological imperative arising from the use of nuclear power. It is an imperative that arose from other causes, and nuclear power fits naturally into it.

FEARS ABOUT TERRORISM

Let us now return to the mainstream of major sociological concern over the use of nuclear power—the possibility that it might be misused for non-peaceful purposes. It is a regrettable fact that we live in an increasingly violent society. Our present generation has seen the emergence of a plethora of terrorist groups, each dedicated to one political cause or another and determined to achieve their objectives by violent means. Would it be prudent to introduce a major technology capable of abuse into a society with such a political backdrop?

In what way could terrorists make use of our nuclear technology to threaten society? How serious a threat can they be to us? What steps can we take to protect ourselves? There are two types of things terrorists could do. They could attempt to disperse, or threaten to disperse, radioactive material into the environment, or they could attempt to steal some fissile material with a view to making and exploding, or threatening to explode, an atomic bomb.

Let us look at a few examples of the kinds of things a terrorist group might try to do. They might try to take over a nuclear power station by force, and then try to make it malfunction in such a way as to cause serious harm to the public. One has to ask if it is credible for them to take and hold the station for a sufficient period of time to do that. The plant is designed to shut itself down automatically if there is a malfunction, so it is not easy to make it go wrong, even if one had a large group of people, each of them familiar with its detailed workings. Also, the time delay between such malfunctions and a potentially dangerous situation arising is quite long—very long, in fact, compared with the time it would take the security forces to recapture the station. Any such action would certainly hit the headlines in all the national papers and TV channels but the chance of any harm coming to anyone apart from the terrorists themselves is remote. There are surely easier ways for any group to further its cause.

Perhaps they might try to steal some highly radioactive material, such as spent fuel, with a view to dispersing it in the water supply or into the atmosphere. Here the terrorists face two fundamental problems. The first is that the amount of material they would have to steal in concentrated form, so that when dispersed it might have some chance of actually harming anyone, is such that unless they had the sophisticated equipment necessary to handle it, it would kill them before they could do anything with it. The second problem is that when dispersed any public harm they might cause (and there would probably be none) would not be immediately obvious—it might lead to an enhanced cancer rate in the local community over the next twenty or thirty years. And even that would only be discernible statistically after a very careful epidemiological study. Hardly the stuff out of which terrorist threats are made.

Highly radioactive material, in the form of spent fuel rods on their way from a nuclear power station to the reprocessing plant at Sellafield, is frequently transported about the country by rail and by road. Could not a terrorist group attempt to hijack such a shipment? As we saw earlier, spent fuel is transported in massive containers weighing up to ninety tons. If the terrorists were to attempt to take off the lid and try to remove the rods they would rapidly receive a lethal dose of radiation. If they were to make off with the container complete with its transport vehicle, the prospect of them making their getaway down the M6—at twelve miles per hour, which is the speed that these transport vehicles move with such a heavy load—with the police in hot pursuit does tend to degenerate the scenario into one of a rather inferior West-End farce.

Suppose they were to try to blow up the container *in situ*? In order to have any impact on such a massive object they would need a rather powerful weapon such as an anti-tank rocket launcher. With such a weapon they might manage to pierce the vessel, and if they did this in a built-up area of high population density this might result in the death of about ten members of the public. Make no mistake about it: rocket launchers are dangerous weapons. In the hands of a determined terrorist group they could be made to inflict considerable harm. For example they could be fired at a crowded pub, cinema or bingo hall, or even a large block of flats. Or if the terrorists wished to inflict harm indirectly, they could aim the rocket launcher at a tanker containing chlorine gas or petrol. Then they might produce death tolls in the hundreds or even thousands.

With this type of terrorist threat a pattern is clearly emerging, and it is a very disturbing one. A highly developed industrial society such as ours contains a wide range of technologies that are potential targets for a terrorist group intent on inflicting harm on the public. One could produce a long and frightening list of such targets—a list on which the nuclear industry would feature a long way from the top in terms of potential harm. It is an aspect of our modern society about which we should rightly be concerned. It would be irresponsible, however, to so mesmerise ourselves with thoughts of a possible threat to our nuclear technology that we ignore, and fail to protect ourselves from, the more likely and much more devastating threats to the other components of our industrial society.

What about the other type of terrorist threat that we mentioned, the possible theft of fissile material with a view to making some sort of nuclear weapon? This surely is a threat very different in nature to those that we have so far discussed.

There are two thermally fissile materials in the world—uranium-235 and plutonium-239. In order even to start thinking about making a nuclear weapon, a terrorist would need more than a critical mass of one or other of these isotopes, in metallic form at high purity. Neither of them exists in the civil nuclear industry in this form with the purity needed. The question to ask, then, is are there any materials handled by the industry from which the fissile isotopes might be extracted to sufficient purity?

Let us first consider uranium-235. The highest concentration it reaches is about 2–3% in the fuel elements of the AGR or the PWR. It is mixed with the non-fissile isotope, uranium-238. In order to achieve a separation of these isotopes, a large plant covering several acres is required. A terrorist group is unlikely to have such a plant and even if it did, or had access to one in some other country, why should it go to the bother of stealing reactor fuel for its input material? Natural uranium exists as ore in many parts of the world and has a uranium-235 concentration of 0·7%. Given that one has a large isotope separation plant it is not much more difficult to extract uranium-235 at 90% concentration starting with naturally occurring and easily obtainable ore than it is by starting from reactor fuel elements. The scenario, therefore, of a terrorist group manufacturing a nuclear weapon using the uranium-235 route is not a credible one.

What about plutonium? Plutonium exists in spent fuel rods at a concentration of about 1%. Spent fuel rods, however, are intensely radioactive because of the fission products they contain and so cannot be handled in such a way as to avoid lethal doses to the operators without the use of elaborate and specialised equipment which is unlikely to be available to people unauthorised to handle it. Unused fast reactor fuel elements, however, are a different matter. They contain about 20% plutonium and are only mildly radioactive and can be handled without such excessive precautions. The other 80% is again the non-fissile isotope of uranium, uranium-238. To separate two different chemical elements such as plutonium and uranium in the form of ceramic pellets would be difficult though less so than separating two isotopes of the same element.

But then there is a snag. Reactor-grade plutonium is not pure plutonium-239. It contains considerable amounts of the higher, non-fissile isotopes plutonium-240 and 241. While these isotopes have little effect on the performance of plutonium as a reactor fuel, the effect they have on its performance as a weapons material is dramatic. If it were so dramatic as to prevent absolutely its use as a weapons material that would be the end of the matter. Unfortunately, it is almost, but not quite, as dramatic as that. Given extreme care and skill, it is just about possible to make reactor-grade plutonium explode. Granted it would be a very unreliable bomb—it is much more likely not to work than to do so—and its explosive power would be very much inferior to that of a bomb made with weapons-grade plutonium (that is pure plutonium-239), but some sort of a bomb it might be.

Not that it is all that easy to make a bomb even given the right materials, as some of the states who have done so have found to their cost. It often took several attempts over several years to get it right, even with the full support of the state behind them. The prospect of a small group of terrorists achieving success in some small ill-equipped laboratory does tend to stretch the imagination to the limits of credulity. That they should be able to do so using the wrong material—that is reactor-grade rather than weapons-grade plutonium—pushes one beyond that limit. It might make a good plot for a James Bond film, but like much of that delightfully romantic fictional genre, its connection with reality is rather marginal. One does hear reports from time to time of schoolboys who have in principle designed a nuclear bomb. No doubt such resourceful youngsters could in principle also design themselves a space rocket to take them to the Moon. It is actually doing it in practice that is the difficult bit.

If I were a terrorist hell-bent on nuclear blackmail, I would steal a fully armed bomb or warhead, of which there are some tens of thousands in the military arsenals of the world. Not that they are all that easy to get at, of course, but as long as the nations of the world feel it necessary to maintain such weapons of destruction, it is a threat than can never be completely ruled out. But stealing material from the civil nuclear industry—and unsuitable material at that—would to my supposedly terrorist mind seem to be a non-starter.

There is nevertheless a reason why terrorists might attempt to steal such material. Their chief aim is not killing innocent people—though they are not averse to doing so incidentally—it is rather to draw attention to their cause. What more dramatic way is there of doing that than by the theft of some plutonium, or some material containing some plutonium? It might well be the wrong kind for making bombs and there could well be very little they could actually do with it in practice to harm people, but everyone has heard of plutonium and the world's press would have a field day. Terrorists would have a much more difficult job attracting attention by stealing some obscure chemical with an unpronounceable name, even though they could get hold of it more easily and could actually do considerable public harm with it if they chose.

There is just one more, but very fundamental, aspect we must mention before leaving the subject of terrorism. Is it healthy for a society to forego the advantages to be gained from civil nuclear power, or indeed any other technology, because it is afraid of a minute fraction of its population who might attempt to misuse it for harm? If we were to succumb to that would it not mean that the terrorists had already won?

THREATS TO CIVIL LIBERTIES

As the nuclear industry handles material that may, if it got into the wrong hands, be used to cause public harm—even though that harm might be more psychological than physical—no reasonable person could object to sensible precautions being taken to prevent this from happening. Stringent security measures are indeed taken at all critical stages. It would, of course, be counter-productive to describe these in detail—lack of knowledge as to what they are is all part of the defence mechanism—but rest assured that they exist.

Fears have sometimes been expressed that the existence of such measures might themselves pose a threat to the civil liberties of ordinary law-abiding citizens. For example, much play has been made on the existence of the Atomic Energy Authority Constabulary (AEAC), a special police force whose task is to guard nuclear establishments of the AEA and BNFL. Not that there is anything new about special police forces, of course: British Rail has one, for example. When in hot pursuit of suspected terrorists the AEAC has powers of search and arrest and, more worrying to some people, where necessary its members can be armed. This should not lead one to believe that every nuclear establishment has the appearance of an armed camp—one needs only to visit such an establishment to dispel that myth. Having access to arms in an emergency does not mean that guards on duty are always armed to the teeth, although during my years at Harwell I did once see one of them carrying a pistol. However, to cast these specials as a 'private army' is a picture too ludicrous for words. As with any special constabulary, it is subject to the control of Parliament through the appropriate Minister, in this case the Secretary of State for Energy.

At the present time we have a nuclear industry, including both power stations in operation and those approaching completion, whose size is capable of supplying about 20% of our national electricity needs. This might eventually grow to the extent that it might supply 40 or perhaps 50% of our needs. As far as I am aware, no one has claimed that, with the nuclear industry at its present level, the measures needed to protect it cause any observable infringement at all of their civil liberties. The cry has always been it's alright now but what about if it grows? I find it hard to believe that if measures needed to protect a 20% component have a negligible effect then those needed at 50% could in any way constitute an intolerable burden, even if they were to rise *pro rata*, which of course they won't. What, might I ask, is the magic percentage between 20 and 50 at which the balance is supposed to be tipped? It is one of those threats that might sound credible if one says it quickly, but the slower one says it the less credible it becomes. It is all part of the fairy tale that many nuclear protestors have persuaded themselves to believe in.

The reality is that we unfortunately live in a world in which terrorism is on the increase, and in which the terrorist has a wide range of targets. Only belatedly, and in many cases not yet at all, are we taking steps to protect ourselves. Anyone who has travelled recently by plane, and has been subjected to the humiliation of the thorough searches that are now accepted as routine, may well wonder to what extent he still has any civil liberties left at all. As we become increasingly aware of the many ways in which terrorists can harm us through our varied technologies, and increasingly insist that measures be taken in an attempt to prevent them, restrictions on our personal liberties can only increase.

It is the existence of terrorism that is a threat to our civil liberties. The existence or otherwise of a civil nuclear industry, in this or in any other country, has, for all practical purposes, a negligible effect on that threat.

PROLIFERATION OF NUCLEAR WEAPONS

At the present time there are six nations who have manufactured and detonated nuclear bombs successfully—USA, USSR, Britain, France, China and India, although India claimed that hers was a 'peaceful nuclear device'. The distinction between such a device and a bomb is not all that clear, although it is believed that the Indians do not yet have an effective delivery system.

The combined destructive capacity of these arsenals is horrifying. A major war fought with such weapons, while it might not entirely eradicate the human species, would certainly obliterate human civilisation. It is amazing that the human race has learned to live under such an unique and unprecedented threat. I have yet to meet anyone who actually likes nuclear weapons and I do not know of any government that is actually proud of the fact that is has a nuclear weapons capability. The motivation for the development and maintenance of these arsenals rests on a basic human emotion—fear, arising from mutual distrust.

That same emotion could well motivate other nations to acquire nuclear weapons. South Africa, Israel and Pakistan are believed to have already started down that road. Others could well follow suit. The prospect of a world in which ever-increasing numbers of countries become armed with nuclear weapons is a frightening one, and can only bring nearer the day of the final Armageddon. But we should be aware of one of the fundamental aspects of the situation. That is that each of the states possessing nuclear weapons developed them independently without any help from anyone else. What they have done independently, others can do also.

An important question to ask is to what extent the development of civil nuclear technology will affect the spread of nuclear weapons. After all, the basic science is the same—both depend on the process of nuclear fission. When developing that basic science, however, the two technologies rapidly diverge. For example, to produce weapons-grade uranium one needs to build a plant to enrich the concentration of the 235 isotope from its natural concentration of about 0·7% to a value in excess of 90%. To produce uranium suitable for reactor fuel, one needs an enrichment of no more than 2–3%. The types of plant needed to achieve these two objectives are vastly different in complexity and scale. On the other hand, to produce weapons-grade plutonium one would build a small reactor without any power-producting plant, from which the fuel elements would be extracted after only a few weeks for reprocessing in order to obtain the relatively pure plutonium-239 needed for this purpose. In a large power reactor the fuel would be left in for several years, and the plutonium eventually extracted from it would be heavily contaminated with the higher isotopes of plutonium, which would not affect its performance as reactor fuel but would make it unsuitable as weapons material. One would choose different technological routes, therefore, to develop the basic science of nuclear fission for civil or for military purposes—a fact that has been demonstrated in practice by countries that have achieved either, or both, of those objectives.

There are some similarities with the development of weapons for chemical or bacteriological warfare—weapons which, although not frequently talked about, could be even more devastating in their effects than nuclear weapons. They are based on the basic principles of chemistry or bacteriology—principles whose alternative development has led to considerable benefit to mankind. As with the development of the basic science of nuclear physics, the technologies needed to achieve civil as opposed to military objectives diverge rapidly. These examples illustrate a fundamental truth. Basic science in itself is not evil, but how man chooses to develop it could be.

Collaboration between nations on the development of civil nuclear technology poses a dilemma. On the one hand, there is the fear that the recipient nations might divert the knowledge acquired to develop a weapons programme in parallel. On the other hand, if there were no such collaboration, with all the international safeguards and inspection that goes with it, such nations would be tempted to go it alone and to develop both civil and military nuclear technologies under a cloak of national security. The Non-Proliferation Treaty, under which the states which have nuclear weapons agreed to supply civil nuclear technology to the states which do not have weapons on condition that the latter forswear the development of nuclear weapons, imperfect though it is, is at least a step in the right direction. At least it is slightly better than doing nothing. The major weakness of the Treaty is that sooner or later the states without nuclear weapons are going to ask why if it is morally wrong for them to develop nuclear weapons should it be morally right for the states with nuclear weapons to keep theirs? I cannot see how there can ever be a satisfactory and convincing answer to that question. If the Treaty is to survive in the long term the question of the phasing out of the arsenals of the states with nuclear weapons must eventually be faced. So we have a dilemma and, as with all dilemmas, there is no easy answer. In fact there is a further complication. If there is insufficient development on a world scale of such a major energy source as nuclear power this could possibly lead, in the not too distant future, to an ugly scramble for the remaining reserves of fossil fuel. This also carries the risk of sparking off a major war.

In this very difficult area there is one point which stands

out clearly: the development of civil nuclear power in one country has no effect whatsoever on the development of nuclear weapons in another. We can illustrate that, in a negative way, by a specific example. If Britain were to abandon its civil nuclear power programme, by what mechanism would that prevent South Africa—a country with indigenous supplies of uranium, a corps of trained scientists and an established industrial base—from developing its own nuclear weapons? As far as I can see, there is no such mechanism.

CHAPTER 6

The Great Nuclear Debate

Only comparatively recently has there been any discernible public interest in energy matters in general and nuclear power in particular. During the era of cheap Middle East oil energy supplies appeared to be plentiful and unending and there were more urgent and pressing matters to concern us. All that changed dramatically with the oil crisis of 1973. Energy suddenly became an urgent topic of conversation and public debate. And what had happened to nuclear energy? Was it not being developed to protect us against such shortages? Where was it? Why had it taken so long?

There was a vague recollection in the public consciousness that 'nuclear' equals 'bombs', and rather big and nasty ones at that. Would not the widescale use of civil nuclear power also pose threats to our health and wellbeing? During the years preceding the oil crisis there had grown up a timely awareness of the impact of man's technology on the environment and public health. Was it wise for us to be polluting the planet and plundering its nonrenewable resources on an ever-increasing scale? What would be the long-term effects of continuing in this way? Should we not find out more about the perceived dangers of nuclear power before we go in for it on a grand scale? Such was the developing public mood in the wake of the oil crisis, and thus the Great Nuclear Debate was born.

Many of us who work for organisations with an interest in

nuclear power have entered the nuclear debate with enthusiasm. As we work for such organisations it could be said that we have an obligation to get involved, and of course this is true. But for many of us I believe there is more to it than that—we actually seem to enjoy it. Public speaking is not everybody's cup of tea; indeed the mere thought of it fills some people with horror. So as public interest in the nuclear debate developed, those of us with a taste for it emerged in response, whether or not we were officially involved in public relations work—usually we were not—and began to take part. At first this was done in a rather *ad hoc* sort of way, but before long our approach became much more systematic. For example, each organisation sets up a talks service, produces tape/slide presentations, films, information packs, etc, and arranges seminars at and visits to nuclear establishments. It was the scale of public interest that has made this more systematic approach necessary. UKAEA staff alone give over 500 talks a year, and the other organisations have a similarly large involvement.

In this chapter I want to convey to the reader something of the flavour of the nuclear debate and how it is conducted in practice. To do that I will tend to dwell on my own experiences. However, I don't want to give the impression that I am the only person involved, nor do I wish to detract from the work of my colleagues, many of whom are doing as much if not more than I am. It is just that I know my own experiences best, and this is after all a very personal book.

Sometimes I am invited to address a totally hostile audience. The invitation, in paraphrase, goes something as follows. We are holding a protest meeting against nuclear power and we want someone from the nuclear industry to hang publicly, and we understand that you might be suitable! I cannot, of course, refuse such a gracious invitation, and never do except when it is totally unavoidable. After all, if you are talking to a totally hostile audience you have nothing to lose, and you might actually win. The objective on such occasions, however, is not to try to convert the audience—it is virtually impossible to shake the faith of the emotionally committed—but rather to hold the 'thin red line'. You have to show that you can stand up to several hours of hostile questioning—most of which is designed not to elicit information but rather to try to catch you out, and is usually framed in the 'when did you stop beating your wife' format—without losing your cool. The debate usually continues after the formal end of the meeting in some local hostelry and one eventually staggers to one's bed, bloodied but unbowed, in the small hours of the morning.

One such occasion occurred in Cambridge in 1979. I was invited to speak at a meeting organised by the local branch of the Ecology Party, a party not noted for its support to nuclear power. When I walked into the room I removed my coat, and a short time later my jacket, and then my sweater. When I got to the stage of loosening my tie, I turned to the organisers and asked was it not so that the Ecology Party was strongly in favour of energy conservation, and how was it therefore that they had asked me to address them in a room that was so obviously overheated? This caused them a certain amount of embarrassment, but they pointed out that although the room was indeed heated to an unnecessarily high temperature they had only hired it for the evening and were in no way responsible for the heating system. This sort of points scoring, although it might add a bit of light relief to the proceedings, should not be confused with the more serious aspects of the debate, which we then went on to discuss at some length. However it does illustrate the point once again that energy conservation, like charity, should begin at home. The meeting started at about 8 pm. I spoke for about an hour, followed by questions for about two hours. We then adjourned to the bar of a nearby college—where we continued informally for a further three hours, eventually breaking up (amicably, I am pleased to say) at 2 am.

I had the opportunity of returning to Cambridge a few months later. This time the meeting was arranged by one of the local churches with a balanced platform of speakers under a neutral chairman. This time I managed to get to bed by midnight. One interesting feature of this event was that the organisers asked for a straw poll both at the beginning and at the end of the meeting to see if there had been any significant change in opinion as a result of the debate. The result of the first vote was 46 in favour of nuclear power and 48 against, and the second vote 61 in favour and 51 against. Although I could claim that was a victory due entirely to power of my oratory, it was more likely to have been a random error in the counting of the votes, which in both cases were expressed by a show of hands. One of the speakers said that he was bitterly disappointed with the vote, because usually at such meetings there is a swing against nuclear power of 40%. I must confess this has not been my experience. More realistically, what it did show was that the audience was evenly divided before the debate and was still evenly divided afterwards. This is what one might expect. Anyone who troubles to go out to a meeting on nuclear power on a cold February evening is likely to have strong views on the subject one way or the other, and it would have been surprising if such strong views had been changed significantly after only a few hours' discussion. What did surprise me was that the audience was so evenly balanced. It is part of human nature to be more highly motivated to go out to a public meeting if one is against something than if one is for it.

By no means all invitations to speak on nuclear power are requests to participate in a verbal prize fight. More often than not it is a request to be informed on this important issue, and comes from an organisation or society wishing to include a talk on the subject in their programme. For example, most school sixth forms nowadays have a weekly general studies period in which the whole sixth form gathers together to consider some issue of topical interest. A short talk on nuclear power followed by a discussion fits well into such a period, and not surprisingly, therefore, I receive many invitations to speak to school sixth forms.

These events are always lively and often overrun the allotted time, which can create some havoc with the school timetable. I remember such an occasion at a school in Cornwall. The discussion had already gone way beyond the official time limit, and after the session had been formally closed and most of the students had dispersed, belatedly, to their next classes, a small group studying A-level English gathered round me to carry on asking questions. Some while later a rather distraught English master appeared at the door, seeking to find out what had happened to his A-level English set. 'Oh sir' they said 'we can't come now; we are discussing something of importance to the future of the world!' 'I'm more interested in the more immediate future', he replied 'Your A-level exams are in a few weeks time. Out!' It was most gratifying to see that we had stirred up so much interest, but I do hope that they managed to pass their A-level English exams as well.

Visits to school sixth forms are always stimulating, but on one occasion we were accused of trying to brainwash the nation's schoolchildren. We had been on a visit to South Wales where we had spoken at a number of schools and public meetings, and one local citizen wrote to the Secretary of State for Energy, no less, complaining along these lines. He sent a copy of his letter to the local newspaper (he was a City Councillor) and a further copy to the local newspaper in Huddersfield (where he was a prospective parliamentary candidate). I know the technique well—as a politician I have used it often myself! The reply he got, from one of the junior Ministers at the Department of Energy, quite rightly pointed out that the Act of Parliament setting up the Atomic Energy Authority in 1954 included an obligation to them to educate and inform the public on matters relating to nuclear power, and that visits to schools and, indeed, other bodies were a legitimate part of their discharging that obligation.

As for brainwashing school children, anyone who has teenage children or has run a youth club for many years and still believes that they can brainwash teenagers into believing anything that they do not want to believe must be living in cloud-cuckoo land! Modern teenagers do not give the deference to their elders that we were accustomed to do at that age. They question anything they are told uninhibitedly and often volubly. Many uncomplimentary things have been said about modern education, but at least it has taught the next generation to think for themselves. Opinion has to be based on knowledge, however, and it is the desire to disseminate that knowledge which motivates our visits. As for our complainant from South Wales, I understand that he has now become a Member of Parliament, so all was not lost!

DEPTH OF PUBLIC CONCERN

No one who has attended as many public meetings as I have could fail to have observed the depth of concern that some people have about nuclear power. Another observation—or rather a deduction, because it only emerges after some reflection on many experiences on the public platform—is that the reasons for that concern are not necessarily the same as those that people give. Many arguments are put forward against nuclear power, be it the effects of radiation, the risk of a nuclear accident, the dangers of nuclear waste disposal or whatever, but these are, I believe, the symptoms rather than the basis of public concern. The real reasons are much more deeply rooted and much less easy to articulate, but they are genuine and felt very strongly nevertheless.

I remember one particular incident which illustrates this discrepancy between the stated and the real concern over nuclear power. I was speaking to the sixth form at a school in Cornwall, and as is often the case this led to a lengthy and detailed question and answer session. The bulk of the questions came from one girl, and they ran through the complete list of the points that usually come up. Afterwards she came up to me to thank me for answering all her questions, which she said that I had done to her satisfaction, but nevertheless she was still worried. She had been trying desperately to hang the overcoat of her concern on one or other of the various pegs of the specific issues that come up in the nuclear debate. One by one the pegs had been broken off, but she was still left holding her overcoat. There could be no doubt that she was deeply concerned. Her problem lay in her inability to articulate that concern.

To argue that the basic reasons for concern go deeper than

the specific reasons that are often stated does not necessarily imply any sinister motives in those who protest about nuclear power. Certainly one can make out a case based on geopolitical arguments that nuclear protest could undermine the economy of the West, and put us at a disadvantage with respect to the Communist world with its much larger fossil fuel reserves and in which nuclear power development is progressing unhindered by any public debate. But to extrapolate from that case to dismiss the nuclear protest movement as merely an undercover Communist plot financed by the KGB is a leap in logic for which there is no justification. It may well be that a few protesters are motivated by such political objectives, but to taint all such protesters with the stigma of being undercover agents for a foreign power would be cavalier, naive and untrue.

Some works have been published which have implied, or have been taken to imply, such sinister motives. One could quite reasonably dismiss them out of hand on the grounds that such implications cannot be substantiated, but it might be wise to sound a note of caution at this point. I was having a quiet drink on one occasion with one of the leading lights of Friends of the Earth, with whom I have frequently shared a public platform. It was at a time when one such book had been published, and I casually turned to him and said 'Come on now, come clean. Are you *really* being funded by the KGB?' He took a long draught of his Guinness and then replied 'No, but we are open to offers!' He was joking, of course, but it did start off a worrying train of thought in my mind. Suppose the KGB really was channelling funds into protest groups in the West, and 'laundering' them in such a way as to make them virtually undetectable—how would one find out about it? It is a potential threat that one must be aware of and guard against. It would be a disaster if groups of genuinely concerned people were undermined in this way. Let us remember that the price of freedom is eternal vigilance.

With that timely and well meant warning, let us now turn from such sinister thoughts and try to unearth what it is that is really worrying some people about nuclear power. It is unlikely that we will find a unique solution to the problem, but we can identify a number of significant threads. The first, as was mentioned right at the beginning, is the association in the public consciousness between civil nuclear power and the atomic bomb. It was as a weapon of fearsome destructive power that the world was first introduced to the energy locked up in the nucleus of the atom. The connection between the civil and military uses of this energy is therefore firmly rooted in the corporate subconscious, and no amount of disclaimers can totally eradicate it. It is this association that gives rise to the specific question about nuclear weapons proliferation and the possibility of terrorism. It is also responsible for the public perception that an accident at a civil nuclear installation must lead to devastation on a scale much larger than is possible from other sorts of industrial plant. We dealt with these specific points earlier but that is, I believe, the basic motivation behind them. It is interesting to note the change in public attitude to this association over the two generations that have passed since atomic energy made its first appearance. The previous generation reacted with relief that a force which had demonstrably been used for ill was now being developed for good. The present generation reacts by regarding the use of nuclear power for good as being tainted by its association with its possible use for ill.

Another basic motivation is fear of the unknown. Most other hazards to which man is subject have been with us since time immemorial, things such as fire, poisons and physical impact. These can be detected by the senses, and because they have been around for such a long time, ordinary people perceive that they can comprehend them—or, if not comprehend them, at least they are sufficiently familiar that they can assess the seriousness and magnitude of their effect. But radiation is something different. It is a new hazard, or at least it is perceived as being new, and it is certainly unfamiliar. Its biological effect is less readily understood. It cannot be detected by any of the senses. What is more, the damage it might cause is not immediately apparent, but only appears years or even decades later. This combination of the unfamiliar and the incomprehensible has endowed radiation with a kind of spooky aura in the public mind. You can receive a dose of radiation without knowing about it, and it might not have any visible effect on your health until many years later. No wonder, then, that radiation gives rise to basic primordal fears. It is this deep-rooted fear of the unknown that gives rise to specific questions such as the effects of effluent from nuclear plants either under normal operating or accident conditions, and to worries about the long-term effects arising from the disposal of nuclear waste.

The Commission of Inquiry into the Three Mile Island accident in 1979 came to the conclusion that the major effect on public health was psychological rather than biological. It was the fear of radiation, rather than radiation itself, that was the major cause of harm. It was a conclusion that was hardly surprising. Here was a community of ordinary down-to-earth folk suddenly thrust into the limelight. The world's press descended upon them, quoting phrases like 'the world's worst civil nuclear disaster'. Is it any wonder that they were scared out of their wits? Even many months later the pall of fear still hung over them. Some local citizens even claimed that they could still taste the radiation in the air (you can't, of course, taste radiation, and there wouldn't have been any there, above natural background that is, several months later anyway). Every medical anomaly in the area was automatically attributed to the effects of radiation from the damaged plant, even though the maximum dose that anyone could conceivably have received could not possibly have caused such effects, based on medical experience of people who had received very much larger doses in the past. Compared with all that a cool assessment that the maximum effect was likely to be that there was perhaps a 25% chance of one person eventually dying of cancer, out of the 400 000 odd who would eventually die from cancer from such a population sample anyway, seemed rather tame. One wonders whether there would have been the same amount of gut fear following a similar, or even potentially more dangerous, accident at some other kind of industrial plant.

To some people nuclear power symbolises bigness—big technology, big government and big business. They are articulating a widespread and justified concern that our lives are becoming increasingly dominated by large institutions with large-scale technology at their command. The individual does not matter any more, as he has less and less control over the decisions that affect him. This is the motivation behind those questions on nuclear power concerned with its centralisation into large units. Nuclear power stations after all generate power in large wodges, which then have to be distributed to the user via the National Grid. It is not the kind of technology that a family can build into its home and control itself, like a domestic heating system or a set of solar panels. Nor could it be conveniently built and controlled by a small community.

To be sure, the emergence of nuclear energy as a power source did not cause large institutional and technological units to arise—they were there already—but it does fit neatly into this existing structure. The fear, therefore, is that with the more extensive use of nuclear power this structure will become more firmly established, and some would consider this undesirable. However, as we saw in an earlier chapter, we do live in a complex industrial society which, with all its faults, has made an enormous contribution to the material wellbeing of ordinary people, and one feature of such a society is that some technologies are most conveniently implemented on a large scale. Whether or not it is desirable, or even possible, for such a type of society to continue into the foreseeable future, and, if it is, how the institutions and technologies of such a society can be more effectively and democratically controlled, is a very much wider issue that we shall be returning to in the final chapter.

One cannot doubt the depth of concern of some individuals over nuclear power. Nor can one doubt the widespread interest in the subject amongst people in general. What is doubtful is whether one can combine these two statements to say that there is widespread concern over nuclear power. From my own experience I would say that this is not so. It is very easy for those involved in the anti-nuclear movement to assume that their own deeply held convictions are more widely held than in fact they are. After all, most of my friends vote Liberal, but still we do not have a Liberal government! People naturally tend to associate and establish friendships with others who have views similar to their own. It is a natural but unjustifiable step to then assume that one's own circle of friends forms a representative cross section of society as a whole. Almost certainly it does not.

Many of the invitations I get are to speak at public meetings. These are meetings to which people come who are highly motivated to do so on perhaps a cold and wet evening in preference to staying at home to watch their favourite television programme, rather than talks, say, to school sixth forms where one has largely a captive audience. Often I am invited to speak at a series of such meetings in the same area, and although there is usually a fair sized crowd, it is by and large usually the same people asking the same questions. It might be an interesting intellectual exercise but the later meetings of the series do tend to have all the excitement of an action replay. After a while one does get to know the 'regulars' quite well and a not unfriendly camaraderie often develops with them. It is all very challenging and stimulating but where is the great British public?

There was the time I went to Jersey. I had been told that there was a tremendous amount of concern on Jersey, particularly over what the French were doing on the Cherbourg peninsula. There was a large reprocessing plant at Cap de la Hague, and plans to build a large nuclear power station at Flamanville, a mere twenty miles or so from the island. The islanders would get no benefit from these developments, of course, but they might have to suffer some of the environmental consequences, and they were all up in arms about it, or so I had been told. Certainly there had been a lot of angry correspondence in the local papers to indicate that this was so. The island government had therefore organised a series of fortnightly lectures on nuclear power and its possible environmental impact, to enable local people to find out what was going on and to give them an opportunity to give voice to their concerns. I was invited over to give the first lecture, which was to be a general introduction to the series. It had been very well publicised, but even so less than a hundred people turned up. I gather that a similar number turned up to the later talks, and from the press reports of them that I saw—at least judging by the names of the people who asked questions—it was by and large the same people. A hundred people, out of a total population of 75 000! I thoroughly enjoyed my visit to Jersey, and I did meet and talk at some length with most of the people who were deeply worried about nuclear power, but there was no evidence to suggest that their deep concern was as widespread on the island as I had been led to believe.

The author speaking at a public meeting on Jersey. (© Jersey Evening Post.)

LACK OF PUBLIC KNOWLEDGE

To many people the whole nuclear business is a bit of a mystery. They have a certain sense of unease about it because they perceive that it could be rather nasty, but they confess that they don't really know enough about it to say whether or not this is justified. A lot of the blame for this lack of public knowledge must fall on those of us who work or have worked in the nuclear industry. After all, we are supposed to be the people who know most about it. Most of us are public servants in one form or another, and we do have an obligation to educate the public in nuclear affairs. I would not go so far as to say that we have failed to discharge this obligation, but it is difficult for us to refute the accusation that we have not done so with as much vigour as we might have done.

Indeed, one of my motivations for writing this book was that at the time there were no up-to-date books on nuclear power written for the general reader by anyone with first-hand knowledge of the industry. The only exception was a Ladybird book written some years ago by one of my colleagues from Winfrith. It is a very excellent Ladybird book but it would not claim to cater for the full spectrum of potential readership.

It was not always so. In fact, in the early days of the atomic age—the late 1940s and early 1950s—scientists were scurrying hither and thither giving talks and public lectures to enraptured audiences on this latest wonder of man's inventiveness. Books, pamphlets and articles in newspapers, magazines and journals appeared in profusion. It was a subject which had gripped the public imagination. It appeared that almost everyone wanted to learn about the new miracle of atomic energy, and atomic scientists responded enthusiastically to this widespread public interest.

Many bizarre incidents happened in this flurry of publicity, but none to surpass one experienced by an older colleague of mine at Harwell, recently and unhappily deceased, who had become very intensively involved in this public lecture programme. He had been invited to speak to a group of people

(I believe it was somewhere in Essex) and as is common with such events it had been originally arranged some months before. He had noted the date in his diary, and thought no more about it until the day before the lecture, when he dug out his notes, selected his slides and gathered together various bits and pieces of apparatus to give his well rehearsed demonstration. He arrived at the lecture hall in good time, set up his slide projector and laid out his Geiger counter, etc, on the table, much to the bewilderment of the audience.

He then sat down while the Chairman went through his preamble and introductions, which he did, concluding with the words 'and now it is a pleasure to welcome Mr Walter Busbridge, who has come all the way from Harwell to talk to us about Mozart's string quartets!' With a blinding flash he suddenly remembered. On this occasion he had not been invited to talk about atomic energy, but because of his extramural interest in music, to talk about—yes, he remembered now—Mozart's string quartets. Many speakers under similar circumstances would have run screaming from the platform, but such was his presence of mind and depth of experience as a public speaker that he rose to his feet—slowly, to clear his mind of things atomic—and then delivered a brilliant hour-long lecture on Mozart's string quartets, without of course using his slides or referring to his Geiger counters.

It is a story which, quite frankly, I did not believe when I first heard it—such things surely don't happen in real life. But I managed to have a word with Walter a few months before he died and he confirmed that it was essentially true. This sort of thing is a public speaker's nightmare. It hasn't happened to me—yet!

As the 1950s wore on, and cheap and abundant supplies of oil started to flow in from the Middle East, the public gradually lost interest in energy in general and atomic energy in particular. To those in the nuclear industry, therefore, it seemed that the public education programme was completed, in that the general public had now become familiar with this new technology to the extent that it no longer excited interest. So the scientists retreated to their more natural habitat of the laboratory, there to fulfil the promise of atomic energy about which they had been speaking so volubly, and the public education programme gradually ground to a halt.

It is a pity that it did, but one can see how it happened. There is nothing more soul destroying than stomping up and down the country trying to tell people things that they no longer wish to hear. Earlier I commented on the fact that only 100 out of a total population of 75 000 on the island of Jersey came to the public meetings that were organised there. How many do you think would have turned up if it had been arranged ten years ago? Little more than the speaker and the Chairman, I would guess.

Then some twenty or so years later nuclear power re-emerged as a topical issue in the public consciousness. The speed and vigour with which it did so took many people in the nuclear industry by surprise. Why after all this time should there be all this public concern? In retrospect one can say that this timescale is not all that surprising. A new generation had arisen who had not had the benefit of all that public education in nuclear affairs of twenty years previously, and to them it appeared to be a mysterious and alien technology of which, not surprisingly, they were deeply suspicious. As young people naturally form the vanguard of any protest movement, they adopted nuclear power, or rather opposition to it, as their cause. Had the nuclear education programme continued, albeit in lower gear, we would still no doubt have had our present nuclear debate, but at least it would have started from a firmer factual base.

In response to this renewed public interest the education programme was restarted and is now in full swing. I sincerely hope that, should this interest start to wane, the education programme will nevertheless continue, so that at the very least should future generations wish to protest about nuclear power they can do so on the basis of fact rather than superstition.

There is a delightful story told about an airport, that was receiving a stream of complaints from local people about the noise, the smell and the general inconvenience that was being caused. So the airport staff decided to embark on a public information programme to explain in detail to their neighbours what was going on and why. Leaflets were produced and distributed, public meetings were held and visits to the airport arranged. As a result of this comprehensive exercise, the airport still received the same number of complaints, but at least they were now well informed complaints! So it should be with the nuclear industry. Outside criticism is healthy, but only if it is well informed criticism. Hysterical denigration based on superstition and lack of knowledge might make entertaining television, but in the long run it is a waste of time.

In fact we must distinguish clearly between public education and public persuasion. In this age of high-pressure salesmanship and ubiquitous advertising it is all too easy to confuse the two. It is most certainly the business of those involved in the nuclear industry to tell people what it is all about—that is education. It can legitimately include a judgment as to how nuclear power could fit into the developing energy scene, both nationally and internationally. Certainly it should include a description of how the technology works, and the possible consequences to public health and the environment in comparison with those of other energy sources and within the context of an industrial society. However decisions on the contribution nuclear power should make to future energy needs should be taken by society as a whole when in full possession of the facts. Ideally that means all of us. Less ideally, though probably more practically, it means by our elected representatives.

So if you want to know more about nuclear power, get hold of somebody who works in the industry to explain it to you. It is not 'secret'—at least the civil applications of nuclear energy are not secret. Nor is it too difficult for the ordinary person to understand. Perhaps you might get lost with some of the more esoteric details of the technology, but any reasonably intelligent person can grasp the general principles involved, given the motivation to do so. As with any other area of human activity, one needs a certain basic knowledge of the subject to be able to form an informed opinion.

DIFFERING NATIONAL TECHNIQUES

It is interesting to compare the methods by which the nuclear debate is conducted in different countries. By and large they are determined by national temperament and tradition. For example, the French, with their taste for violent revolution, take to the barricades. They indulge in pitched battles between nuclear protestors and armed police. I suppose they know what they are doing, but as a mere Englishman I am afraid that I can't see the point in it. Meanwhile the government—traditionally the French have a more centralised bureaucratic form of government than we are accustomed to in Britain—carries on in its own sweet way and does what it genuinely thinks is best for the country.

In the United States, the argument is conducted through the law courts, with objectors using a succession of legal devices to hold up the construction and commissioning of nuclear installations. Their system is such that there are a large number of such devices available, such is the tyranny of a written constitution that was framed long before nuclear fission was discovered. In fact it was framed before uranium was discovered! At least it allows time and a forum for the arguments on both sides to be heard, but the whole process does have a tinge of hypocrisy about it. The prime objective of the objectors is to introduce such inordinate delays into the construction timescale of the project that it becomes uncompetitive economically and so falls by default. Relevant arguments do emerge and are discussed in the process, but only coincidentally. There is one very attractive feature of the American scene, however, and that is their Freedom of Information Act. Under this Act, any citizen can demand to see copies of any government report, unless the department concerned can show that national security would be compromised if it were released. To date the Act has worked rather mechanistically—the random publication of such reports as some individual citizens happen to have heard about, some of which will be incomprehensible and perhaps even misleading, on their own is hardly likely to reveal the full picture—but at least it means that more information becomes available than would otherwise be the case. It is certainly a step in the right direction.

The Russians have a very simple attitude towards the nuclear debate—they don't have one! That is hardly surprising. After all, they don't have public debates about anything. It is not necessarily anything to do with their present political regime. It is something that goes back deep into the roots of their history. They do not have, and never have had, a tradition of open public debate implying, as it must do, some criticism of official policy. As a consequence, the average Russian knows virtually nothing at all about the nuclear programme and its possible consequences that his government is engaging on on his behalf.

This lack of information is particularly regrettable when it comes to accidents. After all, we in Britain had a reactor fire at Windscale (Sellafield) in 1957 and the Americans had an accident at Three Mile Island in 1979, so it would not be surprising if the Russians had also had some problems in the development of their own nuclear programme. In fact there are indications that something rather nasty did happen in the industrial region of the southern Urals in the late 1950s. A Russian exile, Dr Zhores Medvedev, wrote an article about it in a British magazine in 1978 and it created quite a stir. The myth has grown up that this was the first we had heard about it in the West, or worse still, that our authorities knew all about it but were keeping it secret. In fact that's not true. It was reported in the national press at the time. For example, if you would care to refer to the *Evening News* of 18 March 1959 you will find an account of it. It is only a short article, but it does contain almost all we know about it even some twenty years later. You will find it on page 15, which gives some indication of the level of public interest in such matters at that time.

That some incident did occur seems fairly certain, not only based on the 'no smoke without fire' principle but also from a study of some articles in the Russian scientific literature published in the 1960s which seem to indicate unnaturally large levels of radioactive contamination in that part of Russia. What happened exactly, however, is far from clear. Such scanty reports that we have are largely based on hearsay and rumour, usually second or third hand, which are often contradictory and of doubtful reliability. They do not even tie down the date to better than a five-year interval. It might possibly have been a chemical explosion at a liquid waste store—it is known that the Russians were using a rather unsatisfactory method of reprocessing at the time, which they have since abandoned and which we have never used in the West. It might have been a reactor fire like the one we had at Windscale (Sellafield). It might even have been due to the accidental explosion of a nuclear weapon, or a weapons test that produced a larger effect than they expected. We can never know for sure unless the Russian authorities come clean and tell us. Until now they have always flatly denied that any such accident ever happened at all.

As I said, that is most regrettable. It is only by a detailed inquiry into such accidents followed by the publication of a full report that we can hope to learn from our own, and other people's, mistakes and hence improve our practices in the future. That applies to other technologies as well, of course. So if the Russians did have a serious accident in the Urals in the late 1950s I would implore them to produce a report on it. No one wishes to mock or point the finger. Rather, it is a way in which we can learn from each other. After all, one presumes that the Russians have studied and benefited from the reports published following the Windscale (Sellafield) fire and the accident at Three Mile Island.

Having considered how they go about things in other countries, how do we conduct the nuclear debate in Britain? Well, our forte is the Public Inquiry. The best known one was that held in 1977 on the proposed extension of the reprocessing facilities at Windscale (Sellafield). It was an open-ended investigation which went on for a hundred days, by which time everyone who had something to say had said it and everything that needed to be said had been said; the Inspector went off to consider the arguments and write his report. It was a very

British affair. Even those with diametrically opposed views treated each other with the utmost courtesy. During the intervals between the sessions, the opponents and proponents even played football against each other! It just could not have happened anywhere else. When the Inquiry was over, everyone agreed that it had been a good one. It had been scrupulously fair and everyone had had their say.

The spirit of bonhomie evaporated rapidly, however, once the report was published. In a nutshell, it said 'Yes, go ahead'. Many of the opponents of the scheme had spent a lot of time and effort in marshalling and presenting their arguments. To find that their case had effectively been given nought out of ten was seen by them as a slap in the face. Their attitude towards the Inquiry in general and the Inspector in particular hardened noticeably. 'We were robbed!' went up the cry. Foreigners should not make the fatal assumption that because we British are good sportsmen we are also good losers. The natural reaction when we lose is that we should (*a*) shoot the referee or (*b*) change the rules. As the former is not really our custom, let us look at the latter. We are, after all, a pragmatic race and we use the device of the Public Inquiry to examine all sorts of issues, not just nuclear ones such as those proposed for the PWR and the fast reactor, but also for new coal-mining developments, the siting of motorways and airports, etc. So it is important that we get the format right. Inquiries can only be really effective if everybody involved agrees that they are.

One complaint concerns the financial disparity between the two sides of the argument. An Inquiry is set up to investigate some scheme proposed by a large corporation—either public or private—which of course has more than adequate financial resources at its disposal to hire the best Counsel available and to give him all the back-up he needs to present their case in the best possible light. The objectors to the scheme, on the other hand, are usually individuals or small groups of individuals, often formed *ad hoc*, who have no such financial resources. They have to scrape together as much money as they can either from private donations, running jumble sales, etc, or by digging into their own pockets. The lack of financial balance is all too obvious and it offends against the British sense of justice and fair play. Such an imbalance would not be accepted in a criminal court, and for many years we have had a legal aid scheme available to help to correct it. Could we not devise a similar sort of scheme for Public Inquiries?

The case for financial support for objections at a Public Inquiry is in principle a strong one. The problem is how it would operate in practice. A Public Inquiry, after all, is not the same as a criminal court. In the latter there are two well defined opposing parties—the prosecution and the defence—and so it is a relatively straightforward matter to ensure some sort of financial parity between them. But in the former anyone who has anything relevant to say—either for or against the project under consideration—is free to do so. How then does one give financial support to this plethora of contributors, from large organizations to individuals? How much does one give them, and in what relative proportion? The more one thinks about it in detail the more difficult it appears, and the less likely it is that one can come up with an acceptable formula. I am not saying that it is impossible to do so, but it is an extraordinarily complex problem.

However, it does not appear—certainly in the case of the Windscale (Sellafield) Inquiry—that lack of financial resources was the limiting factor in preventing objectors to a scheme from putting forward a coherent and credible case. To be sure, the Windscale objectors had difficulty in raising enough money to present their case, but raise it they did, and I have not heard any claims that pertinent arguments were not presented to the Inquiry because of lack of funds. What ruined the opposition case at Windscale was lack of coordination. Some very reputable and responsible environmental groups went to a lot of trouble to prepare their case and brief Counsel to present it for them. They must have wept tears of anguish to have their pitch queered by a succession of what one could only describe as nutters bringing up a series of ill considered and demonstrably unsound objections. These could only reflect badly on their own case, and damage their credibility in the eyes of the Inspector.

Can anything be done to improve future Public Inquiries? The present system is for the responsible government Minister to appoint an Inspector, usually a High Court Judge, and two Assessors with acknowledged professional expertise in the field under consideration to assist and advise him. Why not extend the system so that the Minister also appoints, and of course pays for, two QCs, one to marshall the arguments and present the case for, the other the case against the project? Any individual would, of course, still be free to make a submission but it would go through the relevant Counsel who would coordinate it within his overall strategy. The suggestion has the advantages of being manifestly evenhanded and also clear and straightforward to operate. It should also lead to a more rational and coherent presentation of the arguments on both sides. It will also go a long way towards solving the problem of financial support. By far the largest item of expenditure opponents find when funding their case is the cost of hiring a QC of high calibre.

The Public Inquiry is but the tip of the iceberg of the nuclear debate in Britain. It is the formal, set-piece debate, and is usually about a specific issue. A decision has to be taken whether or not to start a specific project, be it a nuclear power station or a new reprocessing plant or whatever, and the Public Inquiry is part of the procedure to be gone through before the Minister makes his decision one way or the other. Although more general points do arise, most of the argument is constrained to be directly relevant to the project under consideration. Inevitably the Inquiry is adversorial in character, for at the end of the day the answer has to be 'yes' or 'no'.

The bulk of the nuclear debate is carried out less formally at a large number of small meetings up and down the country. As no specific proposal is to be considered these are much more wide-ranging than a Public Inquiry, and as no decision is required to be taken they can be much less adversorial in character, the object being to disseminate information on the subject, to try to understand the differing views expressed and, where possible, to find common ground. It is not uncommon for a speaker to stand up after another with whose views he has been billed by the organisers to be diametrically opposed and say that he agrees with 80% of what his 'opponent' has said.

On one occasion I was taking part in such a debate on local television. Early on in the proceedings I happened to make the statement that a reactor cannot explode like an atomic bomb. Licking his lips in anticipation of the blood of controversy, the Chairman turned to one of the other participants, who was a member of Friends of the Earth, and asked him what he thought about that. 'Oh, yes I agree' he replied. The Chairman's jaw sank noticeably. 'What on earth are we going to talk about for the next hour?' one could almost hear him thinking. He needn't have worried. We found plenty.

Usually, though there are regrettable exceptions, these debates are good natured. One reason for this is that one frequently finds oneself sharing a platform with the same speakers, and so a certain rapport is established, strengthened of course by a few jars together afterwards in some local hostelry, where the debate continues even more informally, often continuing into incoherence. As a result one gets to know one's opponents, and their views, very well. Indeed there was one such occasion when, for the fourth time in as many weeks, I was having a post-meeting drink with an anti-nuclear speaker. I suggested to him—I forget after how many pints—that if we were to find ourselves sharing a platform yet again, it might make life interesting if I were to give his talk and he give mine.

Nuclear debaters from other lands might find all this fraternisation rather peculiar, but then I'm not impressed by the way they conduct their affairs either! Each to his own I suppose.

PUT IT SOMEWHERE ELSE!

Nothing galvanises and unites a local community like a suggestion that something with the adjective nuclear attached to it is going to be put nearby. People have a vague recollection from what they have read in the newspapers or seen on television that there is something rather nasty about it and sure as heck they are not going to have any of that damned nuclear stuff anywhere near them, thank you very much! In their own way, motorways and airports have very much the same effect. It is what is known as the 'Put It Somewhere Else' Syndrome.

However, it is interesting to note that there is very little local concern about existing nuclear facilities. One would not make much headway in trying to stir up anti-nuclear sentiment in the area, say, around Harwell or around Dounreay. One reason for local acceptance, of course, is the employment opportunities that such a facility supplies, and the fact that most of the money earned there will be spent locally, thus providing further jobs in the service industries in the area. So the facility is seen as contributing to the economic wellbeing of the neighbourhood, thus offsetting to some extent any supposed disadvantages that it might have. Another reason for acceptance is familiarity. When people have got used to a certain type of technology in their area they seem to be less apprehensive about it. Thus there is unlikely to be any local opposition to the siting of the Commercial Demonstration Fast Reactor at Dounreay—on the contrary, the local community has sent deputations to Whitehall demanding that it should be—as they have lived quite happily with fast reactors for several decades. But the suggestion that a nuclear power station be built in Cornwall, where they have no such long association with the technology, provokes violent local opposition.

Acceptance through familiarity is a phenomenon which is not, of course, restricted to nuclear power. In South Wales, where they have a long tradition of coal mining, there is public resentment every time an old and uneconomic pit is closed down. On the other hand, in the Vale of Belvoir, where they have no such tradition but where huge deposits of coal have been found, there is strong local opposition to a pit being sunk.

One cannot doubt the strength of local opposition to nuclear power stations, etc, in areas which have not previously had a connection with the nuclear industry. After all, are there not demonstrations and protest marches in these areas that are featured in glorious technicolour with monotonous regularity on television? The perceptive viewer, however, might notice certain strange anomalies. For example, it is often mentioned in passing that busloads of demonstrators had turned up. Why, one might ask, if it was a *local* demonstration was it necessary to bus the demonstrators in from far and wide? Was that the only way the organisers could get a respectably large crowd? Then again, one invariably sees the faces, and hears the voices, of the leading national figures of the anti-nuclear movement. A local nuclear protest attracts protestors from up and down the country like a honey pot attracts wasps. So even if the local people didn't realise that there was something to protest about there is no shortage of well intentioned individuals available to tell them.

In a sense the 'Put it Somewhere Else' Syndrome is an endemic dilemma of our modern industrial society. We all want the goods and services with which it supplies us, but none of us are prepared to put up with the technological hardware necessary to produce them. A nuclear power station, coal mine, motorway, airport or whatever provides benefit to society as a whole. The inconvenience, pollution and hazard—whether real or perceived—has to be borne by the relatively fewer number of people who happen to live near where they are sited. Very few of us are sufficiently public spirited—though we sometimes like to think that we are—to put up with personal inconvenience for the general good of society. To some small extent the problem is self-correcting in that one community has to put up with the nuclear power station, another the large chemical works, and yet another the motorway or airport. The distribution is far from uniform, however, and certainly this argument cuts no ice with local people when a new project is proposed.

There is some compensation in that the local plant will provide local jobs, a factor that might prove to be increasingly important if unemployment continues to run at its present uncomfortably high level. That can at best be only a partial solution to the problem of how one compensates a local community for having to put up with a piece of technological hardware whose benefits will be shared by all. If our present industrial society is to continue it is a problem that we will have to face up to sooner or later, and with increasing public vociferousness it is more likely to be sooner than later; there is no easy instant solution.

A QUESTION OF CREDIBILITY

One complicating feature of the nuclear debate is the diminished status of the expert in public esteem. When I was a lad, an expert speaking within the sphere of his competence did so with authority and was believed implicitly. It is very different today, when the expert is greeted with almost universal derision. In some ways this change in attitude is not a bad thing. The expert most certainly knows more about the subject in which he is an expert than does the man in the street, but equally certainly he does not have access to divine revelation nor is he infallible. It is a healthy sign when experts are subjected to cross-examination, are made to explain what they know so well in simple terms and are asked to define the limits of, and uncertainties in, their knowledge. But a blanket distrust and denigration of the expert serves no useful purpose. After all, most people are expert at something—that is how they earn their living. So make sure you can play the piano before you shoot the pianist!

There are at least three reasons for the decline and fall of the expert in public esteem. The first is the overbearing arrogance of the expert himself. People nowadays resent being treated like simpletons and being told rather patronisingly to leave it to those who understand these things. The second is that expertise itself has become a debased currency. Time was when to be recognised as an expert one had to subject oneself to the discipline of acquiring the necessary qualifications and then follow this by several years' work in the field to accumulate sufficient experience. Nowadays a brief read of a few light pamphlets combined with an innate gift of the gab seem to be all that is required. The first question to put to an expert, therefore, is to ask him to justify his claim to be one. The third reason is that today most of us have access to the expert through the medium of television, where he is questioned by a professional television interviewer who is himself an expert in the technique of television interviewing. The temptation is great—and is often succumbed to—for the interviewer to try to reduce the expert to a slobbering idiot. Given his own expertise in his profession, it is not difficult for him to do that. It increases his kudos with his fellow interviewers, and if it impresses the boss it could lead to his promotion. All good fun I suppose, but it does help to reduce still further the credibility of the technical expert in the eyes of the public.

To be credible, therefore, the expert has to learn the art of humility. He is not a superior type of human being, but merely a very ordinary person who has acquired a fund of knowledge which it is his duty to convey to his fellow human beings. He must also be prepared to justify his claim to be an expert (in so far as that does not compromise his humility—a difficult one, that!). Finally, he must learn the art of communication and be able to keep his cool when faced with persistent hostile questioning. That is a tall order, but we have come to expect more from an expert than that he merely knows his subject.

Nowhere is the denigration of the expert more evident than in the nuclear debate. There is the added dimension that as the expert works or has worked in the nuclear industry he must be 'biased'. How one acquires an expert knowledge of an industry without ever having worked in it is, I must confess, beyond me. There is nothing more dispiriting than having given a good speech to an audience to be met with a glassy eyed response: 'We don't believe you!'

This is a most unnatural situation for a scientist to find himself in. A scientist is more accustomed to speaking to scientific audiences on the research that he is currently engaged in. Although the audience might occasionally challenge an interpretation that he has put on his results, they would never dream of suggesting that he wasn't telling the truth.

The nature of scientific discussion just isn't like that. So he has to learn the art of not only telling the truth but also making it sound like the truth. One has not only to be honest, one has to look honest as well.

Occasionally this can lead to bizarre situations. For example, I was sharing a platform at a public meeting in Truro one evening with an anti-nuclear speaker. After the meeting a lady from the audience came up to me and said that she was a member of the local branch of Friends of the Earth and had previously only heard the arguments against nuclear power. Having heard me, she said, she was now convinced that nuclear power was a good thing for the country. Naturally I was pleased that my remarks had been so well received, and asked her what it was that I had said in particular that had convinced her. 'Oh, it was nothing like that' she said. 'I looked at you, then I looked at him, and I thought that you looked the more honest!'

So much for the power of rational argument!

CHAPTER 7

The Real Issue

So far I have steadfastly avoided the practice of quoting passages from other people's books and articles. I always find it intensely irritating when reading a book to keep coming across huge chunks from other books. Also, although I might agree with something that someone else has said, usually it doesn't put it quite as I would have done. But I must now make an exception, and quote a passage from an article that appeared in a journal called *The Ecologist*. It is a journal that deals with a variety of environmental issues, and subtitles itself *The Journal of the Post Industrial Age*. It is not that it argues that nuclear power is a work of the devil, it just assumes it is—and that all its readers assume that it is—and goes on from there. With that minor qualification it is an excellent journal. The Harwell Library gets it and I read it avidly each month. Some of you may remember that *The Ecologist* published an article entitled 'Blueprint for Survivial' in 1972. It was a most thought-provoking article and caused quite a stir at the time.

The following passage comes from the Editorial in issue number 9 of Volume 6 (November 1976). After listing various examples of public opposition to nuclear power the author goes on:

> But what is the alternative? How are we to power our expanding industrial society? The answer is that *there is no alternative*. Oil, in the next decades will become increasingly expensive and ever less available—by the end of the century there will be very little left at all. To increase coal pro-

duction on a sufficient scale for it to become the principal source of energy would, amongst other things, involve so much environmental disruption (in the US for instance, it would mean strip-mining vast areas of Montana and Arkansas etc) that it would encounter as much opposition as building nuclear power stations. Besides, all countries do not have coal.

Solar collectors, windmills, water wheels, tidal mills etc can be of use locally but are unlikely to make any significant contribution to the energy requirements of an expanding world industry in the next decades.

In other words: *No nuclear power—no industrial society*. This sounds like the most convincing argument of all for a massive nuclear power programme—but not in the eyes of its many opponents—and this brings us to one of the most important reasons for the opposition to nuclear power: *the nuclear power station has become the symbol of the industrial way of life*—and it is against this that a vast number of people, especially young people, are now reacting.

The italics are not mine but those of the original author, and they highlight the essential points of the passage. They represent a view with which, after several years of following and living with the nuclear debate, I have developed a considerable sympathy.

Nuclear power poses some threat to public health, but no more so than many other technologies of which our industrial society is comprised and a lot less than most. It provides a potential target for terrorist attack, but there are many other non-nuclear targets where the threat is more credible and the possible consequences more terrifying. It can be misused to make weapons of war rather than used for the benefit of mankind, but then again that also applies to many other industries. Chemical technology can be misused to make weapons of chemical warfare, and civil industries based on the science of microbiology can be turned to the manufacture of bacteriological weapons. Even the steel industry is not immune from this argument. One can use steel to make washing machines, refrigerators, cars and railway lines. One could also use it to make guns, tanks and battleships. Why then has the protest movement picked on nuclear power as a target for its vehement attacks?

It all starts to drop into place when one realises that the arguments levelled against nuclear power are in fact equally relevant to industrial society as a whole. For those most deeply involved in opposing nuclear power, the controversy is then not about nuclear power as such, but about our industrial way of life. Are its material benefits sufficient to outweigh its disadvantages—such things as the depletion of natural resources, pollution and its threat to public health, centralisation of power into the hands of the few, and the improved efficiency with which it enables men to make war on each other? That is what the debate is really about. Nuclear power is just a convenient symbol on which it can be focused. This debate on the future of our industrial society is a vital one, but let us realise that it is just that and dispense with the symbolism that is obscuring the real issues.

For the last two centuries or so we have been living in an expanding industrial society. There is no doubt that over that period the wellbeing of the ordinary person has improved progressively and dramatically. It is not obvious, however, that continued industrial expansion will continue to improve human wellbeing—it might do so but it is not obvious that it will do so. The current conventional wisdom assumes that it will, based on the clear correlation between economic growth and improved living standards in the past. But it is just that—it is an assumption, not a law of nature.

There are two factors which could suggest that this assumption might prove to be invalid. One is what we might call the saturation effect. If you were to double the wealth of a peasant living at the subsistence level the effect on his standard of living would be enormous. But if you were to double the wealth of a millionaire it would have hardly any effect at all on his standard of living. The increase in industrial production necessary to double the wealth of a peasant is trivial, but that necessary to double the wealth of a millionaire would be enormous.

If I were to add up what I am worth in financial terms it would probably come to some tens of thousands of pounds. A very rich man would probably be worth some tens of millions of pounds, that is, he is something like a thousand times wealthier than I am. But he does not eat a thousand times as much food, nor does he live in a house which is a thousand times better. So, although he has a higher standard of living than I do, it is not a thousand times higher.

The disbenefits of industrial activity such as pollution and resource depletion are roughly in proportion to the level of that activity. Not exactly proportional, because one does with time learn to handle pollution more carefully and to use resources more efficiently. However, the benefits of increased industrial activity become progressively less and less effective. So one ends up in a classic situation of diminishing returns.

These arguments, of course, apply only to better off citizens living in industrialised countries such as Britain, Western Europe and the USA. They do not apply to the developing world, which comprises the bulk of the world's population. There, living standards are deplorably low, and considerable economic growth is necessary to improve them. And that raises a very serious moral dilemma. I would very much like to see everyone else in the world rising to the same standard of living that I have, but are there enough material resources on the planet to enable them to do so? I suspect that the answer is probably no.

Now that brings us to the second factor which might limit future economic growth, that of resource limitation. We live on a finite planet, with finite material resources. We cannot therefore continue to expand indefinitely without sooner or later running up against the stops. At the present time we are predominantly concerned about energy resources but, as we saw in an earlier chapter, in the longer term, that is for centuries or probably millenia, there are a variety of energy options open to us. It will not be cheap energy as we have known in the past—that was a happy but limited anomaly—but energy there will certainly be. The limiting factor is more likely to be material resources or resources of land on which to grow our food and deposit our waste.

So for a society like our own it is not at all obvious that sustained economic growth, at least as conventionally defined, is either desirable or achievable. There could be a certain amount of growth, but at a reduced—and diminishing—rate compared with what we were used to in the past. Our present economic depression therefore might not be just a routine downswing in the business cycle, but the harbinger of a future era. To hope that the heady days of the 1950s and 1960s will eventually return if only we learn to pull the right economic levers is to rely on what may be a dangerous illusion. It could well be that fundamental changes are taking place in the economic structure of our society. If so, then we must try to understand and come to terms with them.

It is particularly appropriate that we in Britain should be debating the future of industrial society. We were the first nation to enter the industrial age, after all, so it is more than likely that we shall be the first to emerge from the other side. In general there is considerable reluctance to debate such weighty and fundamental matters. We have grown so accustomed to our industrial way of life, which stretches back over many generations, that we tacitly assume that it is the norm and the ultimate state of human society. But it is not necessarily so that it is the norm and that there can be no further human development. It is a legitimate question to ask what follows industrial society. What could it mean, and what will it be like, to live in a post-industrial age?

Our thinking on this topic has been so limited and blinkered that there is a temptation to assume that it means a return to a pre-industrial society. That is one possibility, of course. In fact that is what will happen to us if we don't sort ourselves out. It is what I would call the 'failure scenario'. Be under no illusion, particularly with our very much larger population compared with what it was when we did live in a pre-industrial society—it would be absolutely ghastly.

So a post-industrial society does not necessarily mean a de-industrialised society. We can illustrate that by looking at the previous major step in man's development. We used to live in an agricultural society in which practically everyone who worked did so on the land producing food. Then came the industrial revolution, and gradually the transition was made to an industrial society. Agriculture did not disappear. On the contrary, modern agriculture produces far more food from the land of these islands than our predecessors in the agricultural age ever did. But agriculture no longer employs the bulk of the nation's workforce—in fact, only about 3% of it now works on the land. Most of the remainder now work in manufacturing industry.

Agriculture still survives, and indeed flourishes, in an industrial society. It has become highly productive, as it had to. As people left the land during the industrial revolution to take on more highly paid jobs in the factories, agriculture had to learn to produce more food using fewer people. Increased productivity through automation is now hitting manufacturing industry. Could this be the start of the post-industrial revolution? Unlike the earlier revolution there are no obvious major employers eager to take up the slack. On the contrary, those made redundant through automation are consigned to the twilight world of the dole queue.

This raises a major issue to be resolved when we make our transition to the post-industrial age. If only 3% of the workforce is required to produce most of our food and, let us say, eventually another 3% are required to produce most of the manufactured goods that we need, what will the other 94% of us do to earn a living? Will we work in the service industries? Will we all work shorter hours, and for fewer years of our working lives, spending the bulk of our time on leisure pursuits? Will we produce yet another divided society, with a few rich workers and a vast horde of poor unemployed?

Or will most of us be absorbed in some as yet undefined and undeveloped post-industrial society? If so, then someone had better hurry up and define and develop it. As of now, no one seems to have a clue as to what it might be. In fact, the very name post-industrial society, rather than some more specific title, is but a cloak to cover our ignorance.

Some have even suggested that man is destined for a super-industrial, extraterrestrial future, with the human race spreading throughout the solar system and ultimately the Galaxy itself. At present such thoughts are merely a figment in the imagination of the science fiction writer, but then so were lots of now commonplace things not so long ago. One vital question to be answered, though, is can we from the resources of our home planet generate enough wealth to finance the enormous investment necessary to start such a programme going, or will that prove to be an insurmountable hurdle?

I am well aware of the fact that I have raised far more questions than I have provided answers. That is inevitable, because I don't know the answers. In fact I don't believe that anyone knows the answers and this is hardly surprising because we have as yet scarcely begun to discuss this vitally important issue. Yet discuss it we must, for until we have a clearer idea as to the direction in which the future development of our society is heading, we cannot really make a sound judgment as to how much energy it will require, and specifically within the context of this book, the role that nuclear power will play. To narrow our vision and debate only the nuclear issue in isolation is to fail to grasp the wider context and to lose our sense of perspective.

I have observed in recent months that this broadening of the nuclear debate into one which discusses the wider issues is actually starting to happen. We still have—we always will have—the professional protesters, and no doubt we will continue to have the local objectors opposing specific projects in their own neighbourhoods, but to most experienced nuclear debaters the view is developing that the specifically nuclear debate is played out and that the much wider social issues should now be coming to the fore.

I would even be so bold as to say that for all practical purposes the nuclear debate is over. However, the more general and important debate on the future development of our society is just about to begin.

APPENDIX 1†

World Nuclear Reactor Types

REACTOR TYPES

In a nuclear reactor, the atoms of uranium or plutonium are split by neutrons, releasing energy as heat and more neutrons to sustain a controlled chain reaction. The neutrons are released at high speed and some are absorbed in the uranium-238 atoms present, transforming them into plutonium. There are two basic types of nuclear power reactor: thermal reactors and fast reactors.

THERMAL REACTORS

Thermal reactors have a moderator to slow down the neutrons so that a sufficient number react with the particularly fissile uranium-235. The choice of moderator leads to two families of thermal reactors: graphite-moderated reactors (e.g. Magnox, AGR and HTR), in which the heat is transferred by gas-, and water-moderated reactors such as LWRs (PWR and BWR), SGHWR and CANDU, in which water also transfers the heat.

FAST REACTORS

Fast reactors currently use plutonium fuel, produced initially as a by-product from thermal reactors; the plutonium is sufficiently concentrated for the fission reaction to be sustained by fast neutrons and no moderator is necessary. Without a moderator, the core of the reactor can be compact and have a high heat rating provided an efficient coolant is used to transfer the heat; liquid sodium is used. Excess neutrons are absorbed in uranium-238 both in the core and in a surrounding 'blanket' and 'breed' more plutonium.

† This appendix is reproduced by kind permission of the United Kingdom Atomic Energy Authority.

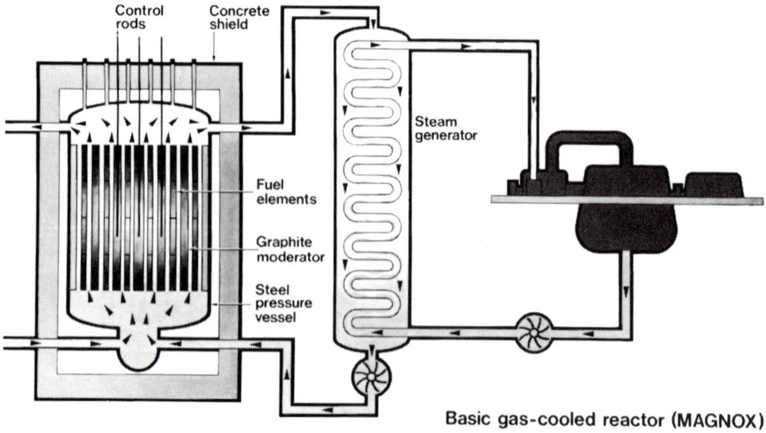

Basic gas-cooled reactor (MAGNOX)

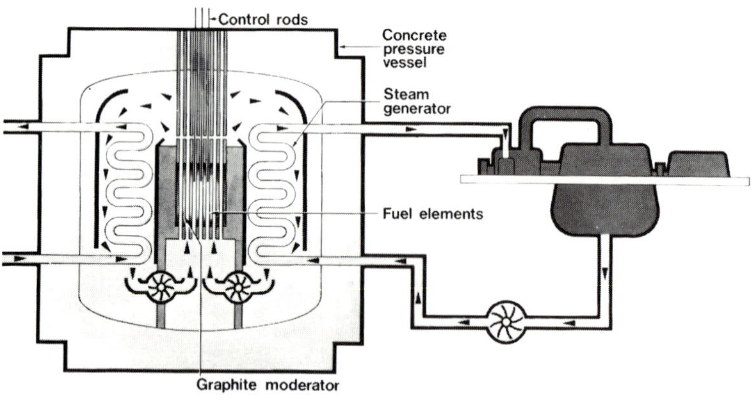

Advanced gas-cooled reactor (AGR)

Magnox Thermal Reactor — Graphite Moderated (UK)

Fuel Uranium metal. To conserve neutrons and allow natural uranium to be used, the fuel is clad in a magnesium alloy (Magnox) with low neutron absorption.
Moderator Graphite.
Heat extraction Carbon dioxide gas is heated by passing it over the fuel in the core and transfers its heat to water in a steam generator; the steam drives a turbine coupled to an electric generator.
Indicative data for a reactor of 600 MW(E) size

Uranium enrichment (% U.235)	0.7% (natural)
Coolant outlet temperature	400°C
Coolant pressure	300 psia
Steam cycle efficiency	31%
Core dimensions	14 m dia. x 8 m high

Advanced Gas-cooled Reactor (AGR) Thermal Reactor — Graphite Moderated (UK)

Fuel Uranium dioxide in stainless steel cladding. The fuel can operate at higher temperatures and heat output rates than Magnox reactor fuel, giving a smaller size of reactor core and a more efficient steam cycle. To achieve these advantages and a greater heat rating, the proportion of U.235 in the fuel has to be increased (enriched uranium).
Moderator Graphite.
Core layout Clusters of fuel elements are joined together end-to-end in a stringer, placed in vertical holes in the graphite.
Heat extraction Carbon dioxide gas is heated by passing over the fuel in the core and transfers its heat to water in a steam generator; the steam drives a turbine coupled to an electric generator.
Indicative data for a reactor of 600 MW(E) size

Uranium enrichment (% U.235)	2.3%
Coolant outlet temperature	650°C
Coolant pressure	600 psia
Steam cycle efficiency	42%
Core dimensions	9.1 m dia. x 8.5 m high

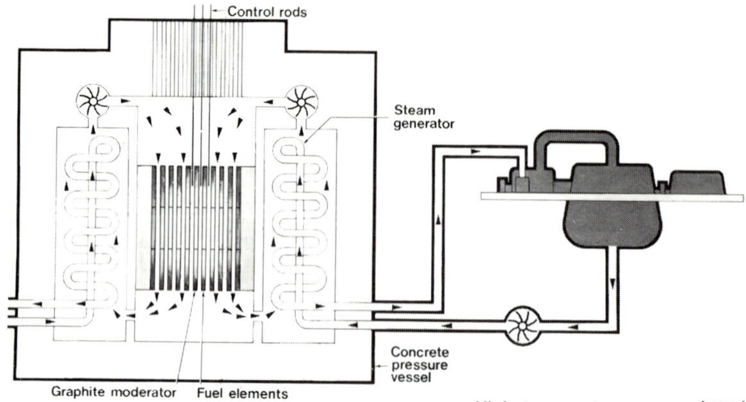

High-temperature reactor (HTR)

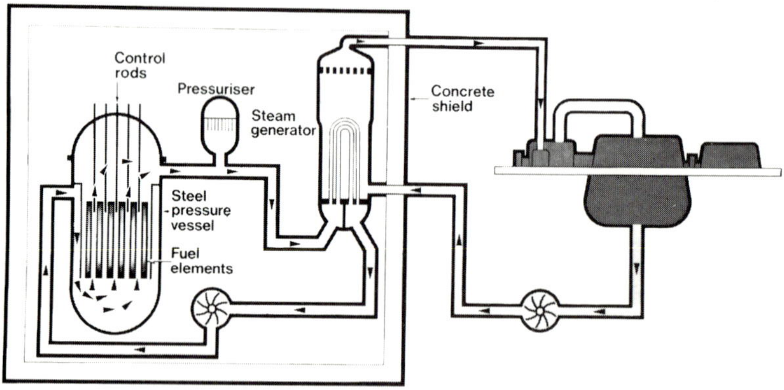

Pressurised water reactor (PWR)

High Temperature Reactor (HTR) Thermal Reactor — Graphite Moderated (UK)

Fuel Small spheres (about 800μm diameter) of uranium dioxide with a coating of silicon carbide, bonded in a graphite matrix and assembled with graphite into a fuel element. The fuel can operate at higher temperatures than metal clad fuel, and a larger proportion can be "burned-up" in each cycle in the reactor.
Moderator Graphite, replaced with the fuel at each fuel change.
Core layout Fuel elements are arranged with vertical coolant passages in the graphite moderator.
Heat extraction Graphite would be corroded by carbon dioxide to an unacceptable extent at HTR temperatures so that Helium gas is heated by passing over the fuel in the core and transfers its heat to water in a steam generator; the steam drives a turbine coupled to an electric generator. Potential developments are to replace the steam generator and turbine by a gas turbine; and to use the heat directly in chemical processes.
Indicative data for a reactor of 1300 MW(E) size

Uranium enrichment (% U.235)	10%
Coolant outlet temperature	720°C
Coolant pressure	715 psia
Steam cycle efficiency	39%
Core dimensions	9.8 m dia. x 6 m high

Pressurised Water Reactor (PWR) Thermal Reactor — Water Moderated (USA)

Fuel Uranium dioxide clad in an alloy of zirconium (Zircaloy).
Moderator Light water (ordinary water, H_2O).
Core layout Fuel pins, arranged in clusters, are placed inside a pressure vessel containing the light water moderator, which also is the coolant.
Heat extraction The light water in the pressure vessel at high pressure is heated by the core. It is pumped to a steam generator where it boils water in a separate circuit; the steam drives a turbine coupled to an electric generator.
Indicative data for a reactor of 700 MW(E) size

Uranium enrichment (% U.235)	3.2%
Coolant outlet temperature	317°C
Coolant pressure	2235 psia
Steam cycle efficiency	32%
Core dimensions	3.0 m dia. x 3.7 m high

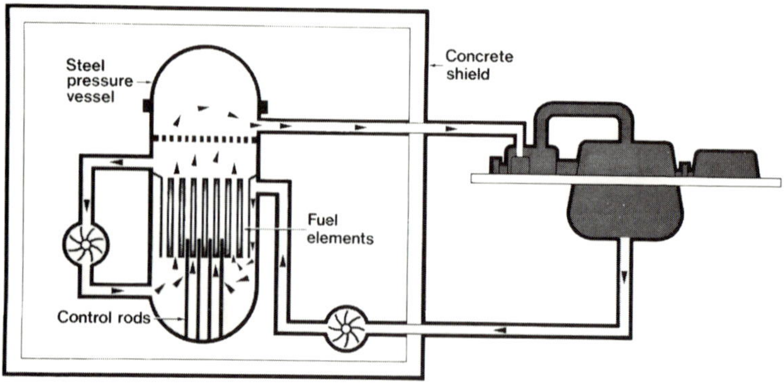

Boiling water reactor (BWR)

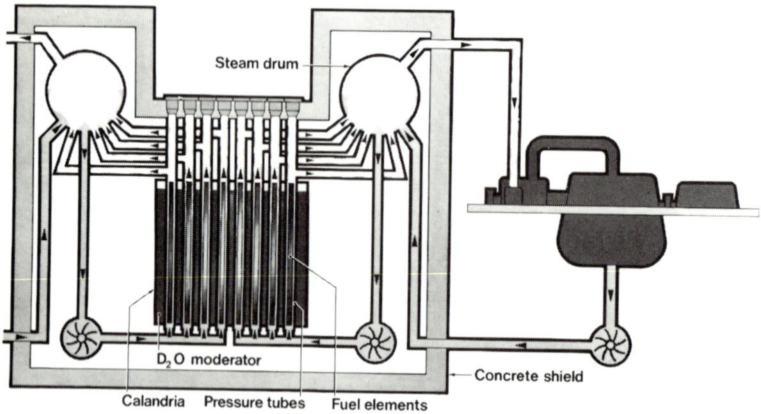

Steam generating heavy water reactor (SGHWR)

Boiling Water Reactor (BWR) Thermal Reactor — Water Moderated (USA)

Fuel Uranium dioxide in Zircaloy cans.
Moderator Light water (ordinary water, H_2O).
Core layout Fuel pins, arranged in clusters, are placed inside a pressure vessel containing the light water moderator, which also is the coolant.
Heat extraction The light water in the pressure vessel is heated by the core and allowed to boil at pressure. The steam from the boiling coolant drives a turbine coupled to an electric generator.
Indicative data for a reactor of 600 MW(E) size

Uranium enrichment (% U.235)	2.6%
Coolant outlet temperature	286°C
Coolant pressure	1050 psia
Steam cycle efficiency	32%
Core dimensions	3.7 m dia. x 3.7 m high

Steam Generating Heavy Water Reactor (SGHWR) Thermal Reactor — Water Moderated (UK)

Fuel Uranium dioxide in Zircaloy cans.
Moderator Heavy water (D_2O).
Core layout Each cluster of fuel elements is in a separate pressure tube; the pressure tubes are in a tank of heavy water. Heavy water is the most efficient moderator and compensates for the neutron absorption in the pressure tubes.
Heat extraction Light water (ordinary water, H_2O) at pressure is heated by passing over the fuel in the pressure tubes and allowed to boil; the steam from the boiling coolant drives a turbine coupled to an electric generator.
Indicative data for a reactor of 600 MW(E) size

Uranium enrichment (% U.235)	2.24%
Coolant outlet temperature	272°C
Coolant pressure	900 psia
Steam cycle efficiency	32%
Core dimensions	6.5 m dia. x 3.7 m high

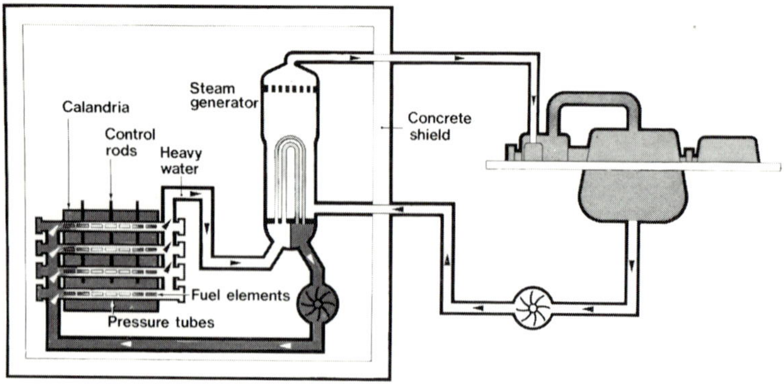

Pressurised heavy water reactor (CANDU)

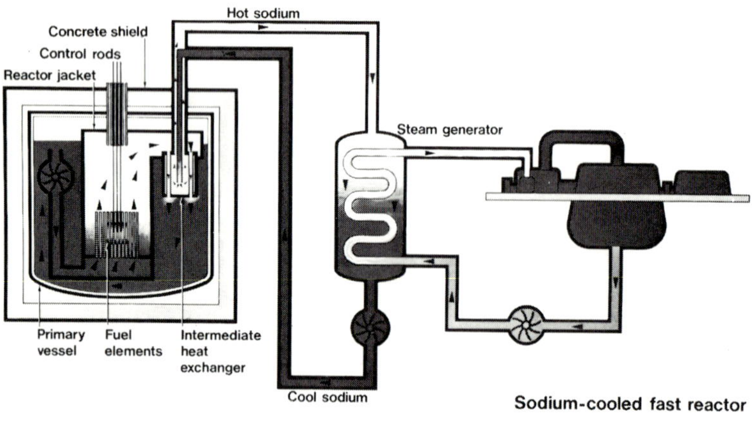

Sodium-cooled fast reactor

CANDU Thermal Reactor — Water Moderated (Canada)

Fuel Uranium dioxide in Zircaloy cans.
Moderator Heavy water (D_2O) which allows natural uranium fuel to be used.
Core layout Each cluster of fuel elements is in a separate pressure tube; the pressure tubes are in a tank of heavy water.
Heat extraction Heavy water (D_2O) at high pressure is heated by passing over the fuel in the pressure tubes. It is pumped to a steam generator where it boils light water (H_2O) in a separate circuit; the steam drives a turbine coupled to an electric generator.
Indicative data for a reactor of 600 MW(E) size

Uranium enrichment (% U.235)	0.7% (natural)
Coolant outlet temperature	305°C
Coolant pressure	1285 psia
Steam cycle efficiency	30%
Core dimensions	7.1 m dia. x 5.9 m high

Fast Reactors — Sodium Cooled (UK. France. USSR)

Fuel A mixture of plutonium and uranium dioxides in stainless steel cans.
Moderator None.
Core layout Assemblies of fuel elements are placed inside a tank containing the liquid sodium coolant. The core is surrounded by a "blanket" of uranium carbide in stainless steel cans.
Heat extraction The sodium is heated by the core and pumped through an intermediate heat exchanger where it heats sodium in a separate secondary circuit. The sodium in the secondary circuit transfers its heat to water in a steam generator; the steam drives a turbine coupled to an electric generator.
Indicative data for a reactor of 1300 MW(E) size

Fuel enrichment (% Pu)	20%
Coolant outlet temperature	620°C
Coolant pressure	5 psia
Steam cycle efficiency	44%
Core dimensions	2.3 m dia. x 1.1 m high
Core and blanket dimensions	3.1 m dia. x 2.1 m high

APPENDIX 2

Bibliography and Sources of Further Information

Selecting a bibliography is a difficult task and often the shortest route to causing offence, so let me start by apologising to those authors who feel that their books ought to appear below but do not. I have used the following selection criteria:

(i) they should be intelligible to the presumed readership of the present book;
(ii) they have been written by authors with first-hand professional knowledge of the subject matter;
(iii) they cover part of the subject matter of the present book and therefore go into more detail on that part;
(iv) they have been read by me and hence can be personally recommended.

Addinall E and Ellington H 1982 *Nuclear Power in Perspective* (London: Kogan Page)

Blair I M, Jones B and Van Horn A (ed.) 1976 *Aspects of Energy Conversion* (Oxford: Pergamon)

Cottrell A 1981 *How Safe is Nuclear Power?* (London: Heinemann)

Farmer F R (ed.) 1977 *Nuclear Reactor Safety* (London: Academic)

Gowing M 1964, 1974 *Britain and Atomic Energy* (3 vols) (London: Macmillan)

Hunt S E 1980 *Fission, Fusion and the Energy Crisis* (Oxford: Pergamon)

Lapp R E 1979 *The Radiation Controversy* (London: Reddy Communications Inc.)

Longstaff M 1980 *Unlocking the Atom: A Hundred Years of Nuclear Energy* (London: Muller)

Martin A and Harbison S A 1979 *An Introduction to Radiation Protection* (2nd edn) (London: Chapman and Hall)

Nuclear or Not? Choices for Our Energy Future 1978 Royal Institution Forum (London: Heinemann)

The Environmental Impact of Nuclear Power 1981 (London: British Nuclear Energy Society)

For further information I would recommend the reader to contact one or other of the bodies with a statutory responsibility in the field, as responsibility has a marvellous way of concentrating the mind. They all produce copious amounts of very readable material at all levels. Particularly recommended is the monthly magazine *ATOM* produced and available free from the UKAEA. The following addresses may prove useful:

Information Services Branch
UK Atomic Energy Authority
11 Charles II Street
London SW1Y 4QP

Information Services Department
British Nuclear Fuels Ltd
Risley
Warrington WA3 6AS

Press and Publicity Office
Central Electricity Generating Board
Sudbury House
15 Newgate Street
London EC1A 7AU

Public Relations Department
The Electricity Council
30 Millbank
London SW1P 4RD

British Nuclear Forum
1 St Albans Street
London SW1Y 4SL

Information Services Department
National Nuclear Corporation Ltd
Whetstone
Leicester LE8 3LH

Public Relations Department
South of Scotland Electricity Board
Cathcart House
Glasgow G44 4BE

National Radiological Protection Board
Didcot
Oxfordshire OX11 0RQ